Adult and Pluripotent Stem Cells

Jürgen Hescheler · Erhard Hofer
Editors

Adult and Pluripotent Stem Cells

Potential for Regenerative Medicine of the Cardiovascular System

Springer

Editors
Jürgen Hescheler
University of Cologne
Germany

Erhard Hofer
Medical University of Vienna
Austria

ISBN 978-94-017-8656-0 ISBN 978-94-017-8657-7 (eBook)
DOI 10.1007/978-94-017-8657-7
Springer Dordrecht Heidelberg New York London

Library of Congress Control Number: 2014933786

Printed on acid-free paper

Springer is part of Springer Science+Business Media (www.springer.com)

General Introduction

There is hardly an area of research developing so quickly and raising so much hopes and promises as stem cell research. Embryonic, fetal, and adult stem cells provide cells with pluri- and multipotency, respectively, and allow deeper insights into mechanisms of self renewal and plasticity. More recently, processes of epigenetic programming and reprogramming to induce pluripotent stem cells (iPS cells) hold the promise for better understanding the differentiation of embryonic stem cells into endo-, ecto- and mesodermal lineages as well as their derivatives. On the other hand, there is great hope that cells derived from stem cells in cell culture can be used for transplantation purposes and regeneration of diseased organs such as the heart (cardiac infarction), brain (Parkinson, neural trauma, stroke) and many other degenerative diseases. This book will concentrate on the perspectives of adult and embryonic stem cells differentiating into unique cell derivatives with a high capacity in regenerative medicine of cardiovascular diseases. The intriguing field of stem cell research under in vitro conditions in the embryoid body model combined with detailed functional characterization of the derived cells has originated in Germany by work of Wobus, Wallukat and Hescheler in 1991 (Differentiation 48:173–182). Since then more than 20 years have passed with a tremendous development of stem cell research. Due to the legal situation many researchers concentrate on fetal or adult stem cells and there are even several clinical studies on the effect of bone marrow derived stem cells after implantation into the infarcted heart. In order to cover this vast area and to bring together the state of the art and expertise a European Consortium INELPY has been formed in which protocols, data, ideas etc. were intensively exchanged to obtain a large surplus (http://cordis.europa.eu/projects/rcn/90957_en.html). This book disseminates the major outcome of this 3.5 years existing group and thus introduces the reader into this fascinating area of research and clinical application, but also incorporates very recent findings interesting for the expert reader. The book is unique as there is as yet no comprehensive overview on adult and embryonic stem cell based therapies. The book also provides high topicality as on the basis of induced pluripotent stem cells (iPS cells) a translation of basic research became possible and the current book will provide a conceptual basis for these future therapies.

The *In*farct *Cel*l Thera*py* (INELPY) Consortium

The INELPY consortium was formed in 2009 to investigate different sources of adult and embryonic stem cells for their potential use to induce and support regenerative processes in ischemic heart disease. The consortium received support from the 7th Framework Programme of the European Commission from 2009 to 2012 (Health-F2-2009-222995: Therapy after myocardial infarction—repair by stem and progenitor cell transfer; see www.infarctcelltherapy.eu). The objectives of the project comprised the evaluation of selected cell preparations, the investigation of potential paracrine factors involved and novel application procedures to improve the effects on tissue repair.

The consortium included four laboratories active in basic and translational molecular stem/progenitor cell research with significant experience in animal infarct models, one group developing novel biomaterial technology and tissue engineering, two cardiology clinicians with experience in clinical cell therapies and two companies with portfolios to develop cell therapies and recombinant growth factors for stem cell growth.

The work performed focused on the one hand on progenitors of vascular endothelial cells and generation of blood vessels, since improved transfusion is the precondition for any repair processes in the damaged heart. On the other hand, the regeneration of heart muscle and heart function was investigated after application of various adult stem/progenitor cells or cardiomyocytes differentiated from embryonic stem cells. This included proof-of-concept studies in small and large animal models. Additional aspects were the improved release of progenitor cells from the bone marrow and the use of specific scaffold materials based on alginate to improve incorporation of the transplanted cells and paracrine effects of growth factors.

Major outcomes of the consortium project showed that a certain type of circulating endothelial progenitor cells, i.e. ECFCs/BOECs, leads to the best long-term engraftment observed for any of the progenitor cells tested, contributed to neovascularization and improved perfusion. As increased perfusion is the precondition for any repair and regrowth of myocardium, these data suggest that intracoronary application of these cells could become a first step therapy to improve perfusion. This could then be followed or combined with therapies designed to replace damaged cardiac tissue by regrowth of heart muscle. In this regard, data from the consortium show that transplantation of stem cells from bone marrow (MAPCs and MSCs) as well as adipose tissue (ADSCs) induce a functional benefit in cardiac tissue despite a very limited degree of long-term engraftment (see contribution by M. Mazo et al.). The mechanisms by which these cells contribute to cardiac repair are presumably to a large extent related to the release of paracrine factors. Some clinical data using BM-derived MSCs have been already obtained by one of the consortium partners (see contribution by J. Kastrup et al.).

Importantly, in regard of proteins/genes involved, we have defined two novel factors which seem to be responsible for at least part of the provascularization functions of ECFCs/BOECs (see contribution by R. Hofer-Warbinek et al.). The data suggest that these factors are responsible for at least part of the positive effects of these cells on perfusion and could be used to support provascularization. Furthermore, the potential of ADSC to contribute to vascularization is described by P.C. Baer and W. Luttmann.

For ex vivo generation of cardiomyocytes significant progress has been achieved (see contribution by B. Krausgrill, M. Halbach and J. Hescheler). Cardiomyocytes have been efficiently generated from embryonic or induced pluripotent stem cells. However, their use in transplantation is still hampered by the low incorporation rate achieved when applied in animal models and the need for correct electrophysiological coupling to the endogenous heart muscle, issues which still have to be improved.

Data using different preparations of biomaterials have further shown that transplantation of cells together with alginate biomaterials improve cell engraftment and function (see contribution by E. Ruvinov and S. Cohen). In addition, significant hopes for future therapies are based on the development of prevascularized patches of heart muscle grown ex vivo from mixtures of stem and/or differentiated cells. In this project we have made significant progress by engineering alginate scaffolds fort his purpose and the application of various electric and magnetic stimulation patterns during cell growth.

In the following contributions of this compendium partners of this consortium review the state-of-the-art of their specific field and summarize their data obtained in the course of the INELPY project. Moreover, S. Janssens and J. Kastrup and collaborators give an overview how the available experimental data from animal models have already resulted in first clinical trials and how these will be further developed in the near future to improve clinical cell therapy of ischemic cardiomyopathy after myocardial infarction.

Contents

Contributors

Miriam Araña Hematology and Cell Therapy, Clínica Universidad de Navarra, University of Navarra and Foundation for Applied Medical Research, Pamplona, Spain

Patrick C. Baer Division of Nephrology, Department of Internal Medicine III, Johann Wolfgang Goethe University, Frankfurt am Main, Germany

Smadar Cohen Avram and Stella Goldstein-Goren Department of Biotechnology Engineering, Ben-Gurion University of the Negev, Beer-Sheva, Israel

Annette Ekblond Cardiology Stem Cell Centre and Cardiac Catheterization Laboratory 2014, The Heart Centre, Rigshospitalet Copenhagen University Hospital and Faculty of Health Sciences, Copenhagen, Denmark

Mandana Haack-Sørensen Cardiology Stem Cell Centre and Cardiac Catheterization Laboratory 2014, The Heart Centre, Rigshospitalet Copenhagen University Hospital and Faculty of Health Sciences, Copenhagen, Denmark

Marcel Halbach Institute of Neurophysiology, University Hospital of Cologne, Cologne, Germany

Jürgen Hescheler Institute of Neurophysiology, University Hospital of Cologne, Cologne, Germany

Erhard Hofer Department of Vascular Biology and Thrombosis Research, Center of Physiology and Pharmacology, Medical University of Vienna, Vienna, Austria

Renate Hofer-Warbinek Department of Vascular Biology and Thrombosis Research, Center of Physiology and Pharmacology, Medical University of Vienna, Vienna, Austria

Ayelet Itzhaki-Alfia Tamman Cardiovascular Research Institute, Leviev Heart Center, Sheba Medical Center, Tel-Hashomer, Israel

Stefan P. Janssens Department of Cardiovascular Diseases, Gasthuisberg University Hospital, KU Leuven, Leuven, Belgium

Jens Kastrup Cardiology Stem Cell Centre and Cardiac Catheterization Laboratory 2014, The Heart Centre, Rigshospitalet Copenhagen University Hospital and Faculty of Health Sciences, Copenhagen, Denmark

Benjamin Krausgrill Institute of Neurophysiology, University Hospital of Cologne, Cologne, Germany

Jonathan Leor Tamman Cardiovascular Research Institute, Leviev Heart Center, Tel-Aviv University, Tel-Hashomer, Israel

Neufeld Cardiac Research Institute, Sackler Faculty of Medicine, Tel-Aviv University, Tel-Aviv, Israel

Werner Luttmann EuroBioSciences GmbH, Friesoythe, Germany

Anders B. Mathiasen Cardiology Stem Cell Centre and Cardiac Catheterization Laboratory 2014, The Heart Centre, Rigshospitalet Copenhagen University Hospital and Faculty of Health Sciences, Copenhagen, Denmark

Manuel Mazo Hematology and Cell Therapy, Clínica Universidad de Navarra, University of Navarra and Foundation for Applied Medical Research, Pamplona, Spain

Beatriz Pelacho Hematology and Cell Therapy, Clínica Universidad de Navarra, University of Navarra and Foundation for Applied Medical Research, Pamplona, Spain

Karl-Heinz Preisegger Institut für morphologische Analytik und Humangenetik, Graz, Austria

Felipe Prosper Hematology and Cell Therapy, Clínica Universidad de Navarra, University of Navarra and Foundation for Applied Medical Research, Pamplona, Spain

Abbas A. Qayyum Cardiology Stem Cell Centre and Cardiac Catheterization Laboratory 2014, The Heart Centre, Rigshospitalet Copenhagen University Hospital and Faculty of Health Sciences, Copenhagen, Denmark

Emil Ruvinov Avram and Stella Goldstein-Goren Department of Biotechnology Engineering, Ben-Gurion University of the Negev, Beer-Sheva, Israel

Caterina Sturtzel Department of Vascular Biology and Thrombosis Research, Center of Physiology and Pharmacology, Medical University of Vienna, Vienna, Austria

Chapter 1
Mesenchymal Stem Cells for Cardiac Repair: Preclinical Models of Disease

Manuel Mazo, Miriam Araña, Beatriz Pelacho and Felipe Prosper

Abstract In recent years, the incredible boost in stem cell research has kindled the expectations of both patients and physicians. Mesenchymal progenitors, owing to their availability, ease of manipulation and therapeutic potential, have become one of the most attractive options for the treatment of a wide range of diseases, from cartilage defects to cardiac disorders. Moreover, their immunomodulatory capacity has opened up their allogenic use, consequently broadening the possibilities for their application. In this review, we will focus on their use in the therapy of animal preclinical models of myocardial infarction, with special focus on their characteristics and their in vitro and in vivo mechanisms of action.

Introduction

In spite of the increasing knowledge of its causes, development and treatment, myocardial infarction (MI) remains the greatest health concern worldwide (Organization WH 2008). Following the ischemic event, the damage inflicted upon cardiac cells rapidly becomes irreversible, leading to massive death and loss not only of muscular but also vascular cells (Mazo et al. 2010b). Although clinical practice has raised the expectancy of surviving an MI and is able to mitigate its progression (White and Chew 2008), the mammalian (and therefore the human) heart lacks a significant capacity to self-heal, leaving a chronically scarred organ with a limited and waning functionality, which will finally lead to heart failure and the dichotomy of transplantation versus death.

During the last 15 years, the boost in stem cell research has ignited the expectations of patients and scientists alike. Although the initial hypothesis was that these plasticity-endowed cells could mend what disease had broken through direct replacement of lost cell populations (Reinecke et al. 2008), an increasing body of evidence points towards their capacity to secrete therapeutic molecules as a potent regenerative tool (Gnecchi et al. 2005). Among the different cell types assayed, mesenchymal stem cells (MSC) are one of the most interesting and intensively

F. Prosper (✉) · M. Mazo · M. Araña · B. Pelacho
Hematology and Cell Therapy, Clínica Universidad de Navarra, University of Navarra and Foundation for Applied Medical Research, Av. Pío XII 36, Pamplona 31008, Spain
e-mail: fprosper@unav.es

J. Hescheler, E. Hofer (eds.), *Adult and Pluripotent Stem Cells*,
DOI 10.1007/978-94-017-8657-7_1, © Springer Science+Business Media Dordrecht 2014

investigated. In the following pages we will revise what characteristics make mesechymal progenitors so appealing for cardiac regeneration as well as discuss the available data from studies in which MSC have been employed to treat animal models of MI.

Origin, Characteristics and Isolation of MSC

In 1974, Friedenstein et al. published one of the first reports on MSC (Friedenstein et al. 1974a, b), demonstrating the clonogenic potential of multipotent cells in the marrow, which they termed colony forming unit-fibroblasts. Since then, our knowledge of these cells has greatly increased, and MSCs are now viewed as a mesoderm-derived population, ubiquitously found in a variety of tissues, most abundantly bone marrow (BM) and adipose tissue (AT), but also in dental pulp, menstrual and cord blood or placenta, among other sites (Mazo et al. 2011; Bianco et al. 2008; Gandia et al. 2008; Hida et al. 2008; Lee et al. 2004; Yen et al. 2005). Nevertheless, their physiological function is still not well established, mainly due to the lack of an appropriate marker for their in vivo identification and the difficulty of performing adequate lineage-tracing experiments. Although the possibility of their residence within the proximity of vessels has been hinted (Cai et al. 2009b), this has not been extensively proved. It is widely assumed, however, that MSCs give support and nurture to other cells, with several studies identifying their role in endogenous tissue repair processes (reviewed in Caplan and Dennis 2006).

This plethora of origins and capacities has impelled standardization, as the Mesenchymal and Tissue Stem Cell Committee of the International Society for Cellular Therapy proposed in 2006, concluding that the minimum requirements to define MSC (Dominici et al. 2006) are: first, cells must be plastic-adherent under currently established culture conditions; second, MSC should express CD73, CD90 and CD105, and lack expression of HLA-DR, CD11b, CD14, CD19, CD34, CD45 or CD79-alpha; finally, they must be able to differentiate to osteoblasts, chondrocytes and adipocytes in vitro. Still, caution must be taken as some reports fail to meet these criteria, and MSC is often employed for "Marrow Stromal Cell", "Mesenchymal Stromal Cell" or "Marrow Stem Cell". Accordingly, a clarification was published in which MSC was defined as "Multipotent Mesenchymal Stromal Cells" (Horwitz et al. 2005), adding the supportive property to the required characteristics (Sacchetti et al. 2007). Nevertheless, tissue of origin gives isolated progenitors a distinct therapeutic potential, making comparisons between different populations difficult. For example, the relative abundance of mesenchymal progenitors in BM versus AT (0.01 %–0.0001 % of cells in the marrow (Pittenger et al. 1999) and 100 to 500 times that number in adipose depots (Mazo et al. 2011) may importantly impact cell growth kinetics and other characteristics, as has been proved at a functional, genomic and proteomic level (Wagner et al. 2007; Kern et al. 2006), suggesting a higher degree of commitment of BM-MSC to chondrogenic and osteogenic lineages than adipose-derived MSC (termed adipose-derived stem cells,

ADSC) (Gimble et al. 2007). Placenta-derived mesenchymal cells share many of the characteristics of MSC, but have been shown also to express embryonic stem cell antigens as SSEA-4, Tra-1-61 or Tra-1-80 (Yen et al. 2005), whereas umbilical cord-MSC (Lee et al. 2004) represent a more juvenile source of cells, not subjected to patient specific constrains as the negative influence exerted by of age or disease (Dimmeler and Leri 2008). Moreover, one of the most appealing characteristics of MSC is their capacity for suppressing the host immune response, but this has only been well established for BM-MSC (Le Blanc et al. 2008). In general however, it is not clear whether the mentioned differences are due to true divergences in the nature of the progenitor populations or to the in vitro processing. The tissue of origin clearly influences the properties of the cells, as already remarked for the case of cord blood. Head-to-head comparisons would allow the solving of this problem, but they remain a scarce investment (Luan et al. 2013; De Ugarte et al. 2003).

The isolation of mesenchymal progenitors is nowadays a straightforward technique, which commonly relies on obtaining an intermediate mixture of progenitors. When the starting sample is BM or blood, a Ficoll gradient centrifugation or similar process is employed to remove most contaminating erythrocytes (Hida et al. 2008; Chang et al. 2008; Mazo et al. 2010a, b), rendering the mononuclear fraction. AT processing requires careful mincing plus an enzymatic (usually collagenase) digestion of the specimen to segregate adipocytes from the stromal vascular fraction (Mazo et al. 2008). In the case of term placenta, after draining of any remaining cord blood, the tissue is similarly minced and digested. In vitro culturing under similar conditions, gives rise to the purified MSC populations, be it from the mononuclear fraction (for BM and cord blood MSC) or from the digested tissue (for AT and placenta). Thus, unless allogenically employed, the use of mesenchymal progenitors implies several weeks of processing prior to their application to the patient. This issue will be further discussed below, as it has fundamental implications for their use in the cardiovascular field.

The Triple Goal of Cardiac Regeneration and MSCs

The reconstruction of the diseased left ventricle (LV) is a final goal with three main aims: (1) the regeneration of the cardiac muscle (Rogers et al. 2011), (2) the formation of a mature, functional vascular network to irrigate it and (3) the return of the damaged tissue to its former geometry (Dixon et al. 2009). Failure in accomplishing any of the three would impair the outcome of the therapy. Stem cells, and mesenchymal progenitors among them, can contribute to these mainly through two mechanisms: certain differentiation to the desired phenotypes and, of more significance, releasing of molecules with therapeutic potential. Further increasing their capacities, mesenchymal cells have also the ability to regulate the immune response, which further widens their possible applications (Fig. 1.1).

Transdifferentiation of MSC to the desired phenotypes has been documented. BM-MSC have shown their capacity to give rise either to endothelial cells (EC)

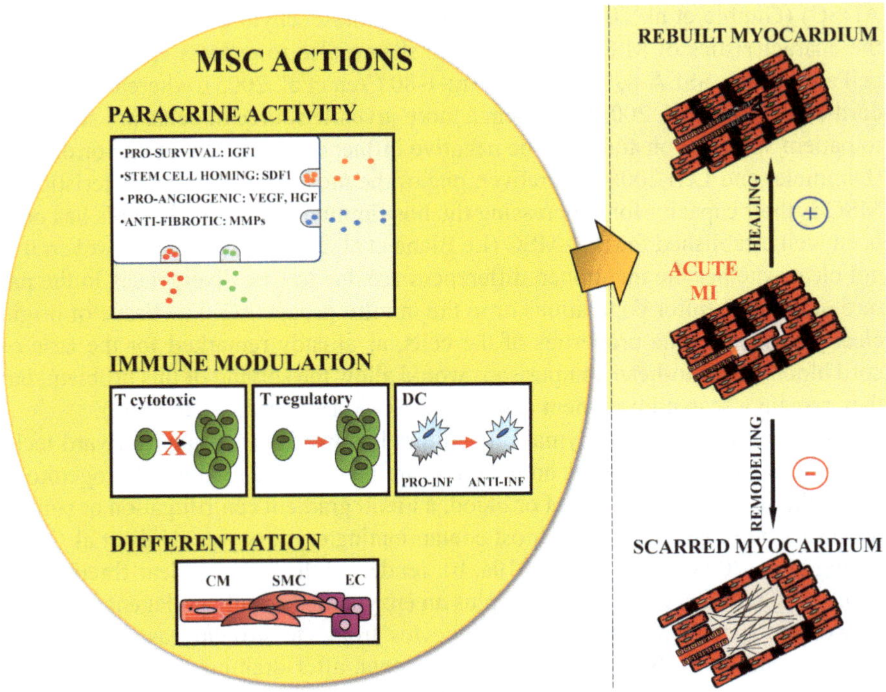

Fig. 1.1 MSC actions on injured myocardium. Mesenchymal progenitors transplanted onto the ischemic myocardium influence infarct healing through three main processes: *i* the secretion of therapeutic molecules (paracrine activity), *ii* the modulation of the immune response, promoting an anti-inflammatory state and *iii* giving rise to differentiated (cardio)vascular cells, encouraging the healing of the damaged tissue and avoiding its transition to a scarred muscle. Abbreviations: *IGF-1* insulin-like growth factor-1, *SDF-1* stromal derived factor-1, *VEGF* vascular endothelial growth factor, *HGF* hepatocyte growth factor, *MMP* matrix metalloproteinase

(Lozito et al. 2009; Liu et al. 2007) or smooth muscle cells (SMC) (Le Blanc et al. 2008), but the obtaining of cardiomyocytes (CM) has been contested due to the immaturity of the resulting cells (Yan et al. 2011; Arminan et al. 2010) or the use of demethylating agents (Xu et al. 2004). Similarly, although ADSC are able to differentiate into vascular cells (Kim et al. 2008; Planat-Benard et al. 2004), the obtaining of CM has been accomplished under treatment with DMSO (van Dijk et al. 2008). Other types of MSC have also been differentiated to CM or CM-like cells, such as menstrual blood-derived MSC (Hida et al. 2008) or umbilical cord-MSC (Chang et al. 2008). In recent years AT has been proposed as a much richer source of stem cells with cardiac and endothelial potential than the BM, yet a cautionary note must be risen as these studies either rely on freshly isolated cell subpopulations or they culture cells under differentiating promoting conditions without true prior mesenchymal enrichment (Leobon et al. 2009; Sengenes et al. 2007), so direct comparison between those and BM-MSC is not possible. To date, though angiogenic potential of MSC is better established and is probably related to the pericytic nature of these cells (Cai et al. 2009b; Jain et al. 2008), their differentiation into cardiac myocytes is still heavily challenged.

However, the suitable exploitation of the differentiation capacity of MSC in MI still presents certain caveats. First, patients receiving stem cell therapy are severely diseased and usually elderly, two factors that have an outstanding impact on stem cell function. For instance, a decrease in the numbers and functionality of circulating endothelial progenitors is directly related to cardiovascular risk and smoking (Kondo et al. 2004; Vasa et al. 2001) and age has also been shown to impair the angiogenic capacity of both BM-MSC and ADSC (Liang et al. 2011; Madonna et al. 2011). Second, cell engraftment is poor, with few cells capable of long-term persistence (reviewed in Haider and Ashraf 2008) in the scarred myocardium. Third, the loss of a significant mass of contractile cells and their associated vessels imposes a burden that is almost impossible to overcome. The myocardium hosts on average 20 million cardiomyocytes (CM) per gram of tissue which, considering a human left ventricle (LV) of about 400 g, means that in cases when an MI damages 25 % of it, about 1 billion contractile cells are lost, as well as other components of the cardiac architecture (fibroblasts, endothelial and smooth muscle cells, etc.) (Robey et al. 2008). Thus all of these factors, coupled to the low rates of cell differentiation achieved even under in vitro controlled conditions, make the rebuilding of the LV through the addition of such a small number of cells a rather naïve strategy.

On the other hand, MSC have been shown to be first-class paracrine secretors, being able to induce a significant effect upon the damaged heart, even when low numbers of cells manage to engraft (Fedak 2008). Angiogenic molecules like vascular endothelial growth factor (VEGF) or hepatic growth factor (HGF) have been shown to be secreted by either BM-MSC or ADSC (Oskowitz et al. 2011; Traktuev et al. 2008; Chen et al. 2008b; Kilroy et al. 2007; Song et al. 2007). Additionally, ADSC have been reported to form and stabilize functional vascular networks when mixed with endothelial progenitors (Traktuev et al. 2009) and Chen et al. showed that BM-MSC were a more potent source of pro-angiogenic cytokines than dermal fibroblasts, exerting a stronger effect upon the recruitment of EC and macrophages and improving wound healing (Chen et al. 2008b). The role of mesenchymal progenitors such as fibrosis-regulators is similarly extensively documented. Conditioned medium from BM-MSC decreases cardiac fibroblast proliferation and modifies the secretory equilibrium towards an anti-fibrotic milieu, with diminished expression of collagen types I and III (Li et al. 2009b; Ohnishi et al. 2007) and increased secretion of matrix metalloproteinases (MMP) 2, 9 and 14 (Mias et al. 2009). Likewise, ADSC produce transforming growth factor (TGF) β1, one of the most potent regulators of fibroblast behavior and fibrosis (Rehman et al. 2004). Beside perfusion promotion and collagen architecture regulation, MSC are able to migrate to the sites of injury due to their expression of the chemokine receptor CXCR4 (Cheng et al. 2008b) and traffic through type I collagen membranes as demonstrated by the expression of five types of MMPs (2, 3 and membrane type-MMPs 1, 2 and 3) (Rogers et al. 2011). These two features would theoretically allow MSC to home to and migrate through the scarred myocardium.

Taken as a whole, all of these exemplify the plethora of paracrine-mediated actions that mesenchymal progenitors can set into motion.

As already noted, immune regulation is one of the most interesting competences of MSC (reviewed in Le Blanc 2006). Marrow-derived MSC inhibit the proliferation of activated T cells and the formation of cytotoxic T cells (Aggarwal and Pittenger 2005), inducing an anti-inflammatory phenotype. This unique capacity significantly broadens the scope of MSC use, as it opens the way to allogenic applications or even to their use as "inducers of tolerance" to other cell types (Bel et al. 2010). However, some concerns have been reported. Huang et al. reported that differentiation reduced the capacity of immunological escape of BM-MSC (Huang et al. 2010b), related to an increase in immunostimulatory molecules MHC-Ia and II and a decrease in the immunosuppressive MHC-Ib. Along similar lines, McIntosh and coworkers reported that ADCS beyond passage 1 (and thus devoid of contaminating differentiated cells) (Mitchell et al. 2006) failed to elicit a response from allogenic T cells (McIntosh et al. 2006), but this attribute may be diminished under inflammatory stimuli, as shown in vitro (Crop et al. 2010). As a consequence, the true ability of MSC to be tolerated by the host's immune system is still debated, and has proved not so significant as initially thought (Puymirat et al. 2009).

Finally, since the groundbreaking report in 2006 by Yamanaka and coworkers (Takahashi and Yamanaka 2006) where induction of pluripotency in adult (stem) cells by overexpression of the pluripotency transcription factors Oct3/4, Sox2, Klf4 and c-Myc was shown, mesenchymal cells have been investigated intensively (Tat et al. 2010; Sun et al. 2009). Advantages like the ease of harvest and culture together with a higher potency than other cell types (e.g.: dermal fibroblasts), has made them good candidates for iPS derivation.

MSC in Preclinical Models of MI

The assessment of stem cells in animal models of MI is a compulsory test before any further step is taken. Results from clinical trials that have already finished have demonstrated the critical impact that not only the treatment, but most importantly the type of disease where cells are applied on, has on the outcome of the study (reviewed in Menasche 2011). Although cardiovascular diseases have been widely modeled, stem cell therapy has been mostly centered on permanent/transient models of myocardial ischemia, with few reports dealing with other settings (de Macedo Braga et al. 2008). However, a distinction between three different models can be made: acute, sub-acute and chronic models of MI.

In the acute form, cells are transplanted within hours of the induction of ischemia. Here, cell therapy must cope with a pro-inflammatory microenvironment and the necrotic/apoptotic molecules released from dying cells (Nian et al. 2004; Mann 1999). Nevertheless, both the presence of homing signals and an anti-fibrotic milieu can be a positive counterbalance (Penn 2009; Cleutjens et al. 1995). The great majority of published reports has relied on this model (Chang et al. 2008; Ii et al. 2011;

Gaebel et al. 2011; Bai et al. 2010, 2011; Dubois et al. 2010; Yang et al. 2009; van der Bogt et al. 2008, 2009; Dixon et al. 2009; Imanishi et al. 2008; Hashemi et al. 2008; Hale et al. 2008; Carr et al. 2008; Cai et al. 2009a; Valina et al. 2007; Li et al. 2007) due to the fact that transplanting cells at the time of coronary occlusion (or minutes/hours after it) subjects animals to only one operation and consequently decreases the mortality associated with the process. With the exception of studies by van der Bogt et al. (van der Bogt et al. 2008, 2009), all specialists have consistently demonstrated that therapy with MSC induces a significant benefit upon cardiac function, which has mainly been attributed to the paracrine activity of the cells that induces an increase in tissue perfusion and a decrease in the size of the scar and collagen content.

In spite of these results, implementation of a therapy based on transplanting mesenchymal progenitors (a population that needs several weeks of in vitro purification) on the acute phase of the disease is not possible unless allogenically employed. Strikingly, far fewer studies have dealt with the opposite situation, the chronic phase (Mazo et al. 2008, 2010a; Song et al. 2010; Lee et al. 2009), where the repair processes take place after the ischemia has been completed: inflammation has receded, the scar has matured and although a new vascular network has been developed, it is disorganized and insufficient (Virag and Murry 2003; Sun et al. 2002). Notwithstanding the great burdens that these conditions impose upon transplanted cells, the abovementioned studies have demonstrated that, as shown for the acute phase, MSCs exert a positive action on cardiac contractility and histology, again mostly due to the secretion of therapeutic molecules and the action on resident cells.

Finally, a third, intermediate setting can also be found: the sub-acute. In this, angiogenic processes are still on course, either through endothelial progenitors (Jujo et al. 2008) or macrophages (Nahrendorf et al. 2007), and the receding of inflammation is coupled to an increase in fibrotic processes. Again, reports in models of sub-acute MI are scarce (Gandia et al. 2008; Hida et al. 2008; Wang et al. 2009a; Amado et al. 2005), but the benefit and mechanisms, as in acute and chronic models, appear to be consistent.

Thus, a fair amount of information is available, supporting the capacity of mesenchymal progenitors to exert a benefit on cardiac function. The mechanistic basis of this is not yet fully explained and it seems to be far more complex than what was expected. However, it is widely accepted that MSC exert their influence upon cardiac contraction not by directly resupplying the tissue with new CM, but through more subtle actions on organ architecture and histology. Some of the abovementioned studies provide crucial hints. Lee and coworkers showed that transplantation of BM-MSC in a rabbit model of chronic MI induced an increase in the concentration of stromal derived factor (SDF) 1, which elicited the chemotaxis of host-derived BM progenitors (CD34+, CD117+, STRO1+) and was related to a functional benefit, a decrease in infarct size and improvement in tissue vascularization (Lee et al. 2009), as similarly shown by Suzuki et al. in a swine model of hibernating myocardium (Suzuki et al. 2011). Li and collaborators demonstrated that the enhanced contractility was accompanied by a preservation in the pro-survival signaling of Akt (Li et al. 2009a), whereas Mias et al. showed that this functional improvement and remodeling reversal in vivo was coupled to a myriad of in vitro anti-fibrotic actions (Mias et al. 2009). The Spinale group monitored the evolution of MMPs and their inhibitors after infusion of BM-MSC in a model of sheep MI, establishing a relationship between the number of transplanted cells and enzymatic levels (Dixon et al. 2009). Resembling what was found in vitro, several reports have associated the pro-angiogenic activity of the cells with the secretion (either directly or host-derived) of potent angiogenic cytokines such as VEGF, HGF or insulin-like growth factor (IGF) 1 among others (Gandia et al. 2008; Li et al. 2007, 2009a; Wang et al. 2009a, b).

The Problems and Their Solutions

Despite the initial optimism, the fire that fueled the field in its beginnings has cooled down. Although positive, results from clinical trials have not been as good as it had been anticipated (reviewed in Menasche 2011) and the desired goal of replacing the scar with a functional myocardium has not yet been achieved. Cell therapy presents certain caveats that need to be improved before the goal of regeneration is achieved. There are two main difficulties: lack of efficient differentiation and low survival/engraftment capacity. As one of the spearheads of cell therapy, MSC are currently undergoing intense investigation to solve both of these issues (Fig. 1.2).

The question of improving cell survival and permanence at the site of injury has been dealt with by two chief approaches: cell modification and tissue engineering. Albeit with certain constraints, genetic manipulation of MSC has undergone profound investigation. Among the many possibilities, the CXCR4/SDF1 axis has been shown to have a capital role in cell trafficking and homing. As a consequence, many reports have exploited it, trying to improve the therapy. Ma et al. investigated the peak of cardiac SDF1 expression after MI in rat, showing that injecting cells at that time point (1 day post-infarction) significantly enhanced cell engraftment and angiogenesis (Ma et al. 2005). Cheng and collaborators injected BM-MSC engineered to overexpress the receptor CXCR4, strengthening cell-homing to the injured tissue (Cheng et al. 2008b). The same group combined BM-MSC peripheral injection with administration of granulocyte colony-stimulating factor (G-CSF). Although G-CSF was able in vitro to increase CXCR4 expression and in vivo to augment cell engraftment, no effect on cardiac function was found (Cheng et al. 2008a). Recently, the group led by Dzau showed that overexpression of the chemokine receptor CCR1, but not CCR2, was associated with improved survival and grafting in a mouse model of MI, as well as functional restoration (Huang et al. 2010a).

Similarly, the issue of cell survival has been dealt with through genetic approaches. Liu et al. engineered BM-MSC to overexpress angiogenin, which conferred an improved resistance to hypoxia and was translated into an increase in cell engraftment and functional/histological recovery (Liu et al. 2008). Adenoviral-transfection of hemeoxygenase-1 resulted in a superior therapeutic effect, mainly through protection from inflammation and apoptosis (Zeng et al. 2008). Akt-overproduction in

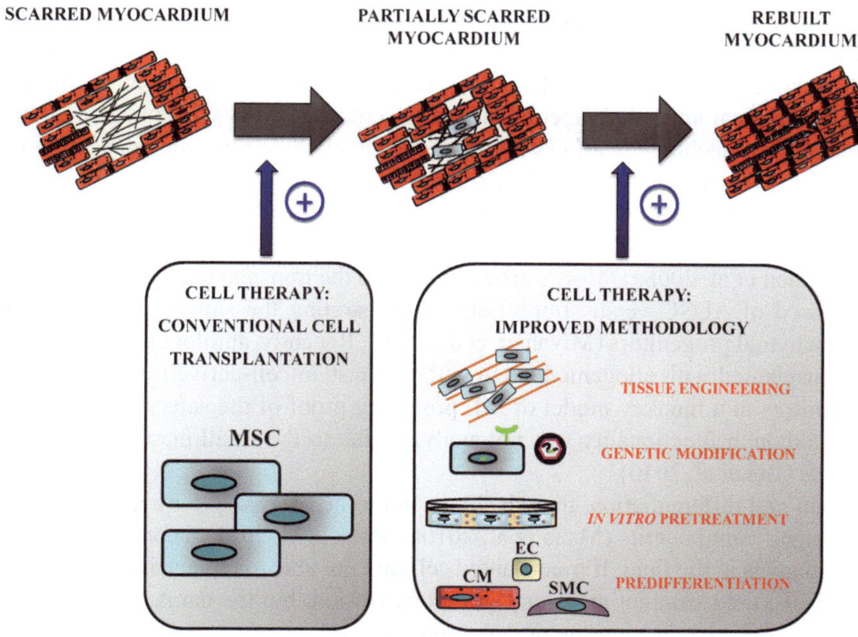

Fig. 1.2 Different strategies for an improved cell therapy of MI. After MI, albeit significantly preventing acute death and transiently increasing contractile capacity, conventional medical treatment does not restore organ architecture. MSC transplantation is able to induce a partial recovery of cardiac function and limit infarct expansion. However, low engraftment and differentiation impair the therapy. Four systems aim at improving this situation and manage total tissue regeneration: tissue engineering, genetic modification of cells, pretreatment with protective molecules and in vitro predifferentiation towards the desired lineages

MSC was able to restore cardiac function 2 weeks post-MI via paracrine actions which included protection from hypoxia-induced apoptosis, release of cytokines and preservation of tissue metabolism (Gnecchi et al. 2005, 2006, 2009). The use of antioxidants, aimed at diminishing injury by reactive oxygen species (ROS), has proved useful, as demonstrated by Song and collaborators who showed that ROS inhibited BM-MSC adherence to the substrate, whereas pretreatment with the ROS scavenger N-acetyl-L-cysteine improved engraftment and precluded the increase of fibrosis and infarct size (Song et al. 2010). Hsp20-mediated protection from ROS also enhanced the therapeutic activity of MSC due to increased expression of Akt and secretion of growth factors (Wang et al. 2009b).

However, genetic modification of stem cells, though possible, implies risks that make it difficult for a treatment to reach the bedside. Bioengineering, on the other hand, makes use of biocompatible materials to improve or direct cell therapy through providing anchorage to transplanted cells as well as mechanically helping the diseased ventricle. Naturally-derived materials have been employed intensively. Porcine small intestine submucose, a decellularized substrate, seeded with BM-MSC, has been used to treat chronic MI in rabbit, resulting in a significant benefit for cardiac function and histology, as well as improved cell retention and migration towards the injured tissue (Tan et al. 2009). Jin and coworkers applied patches of poly (lactide-co-ε-coprolactone) overlaid with BM-MSC on a rat cryoinjury model, where treatment improved functionality and decreased infarct size (Jin et al. 2009). The cell sheet technology allows increasing the thickness of the constructs through stacking. Thus, Chen and coworkers transplanted 5-fold layers of marrow-derived MSC on a rat model of cardiac ischemia, showing that the treatment improved cardiac function along with paracrine secretion of therapeutic molecules by grafted cells (Chen et al. 2008a). Mori's group compared the transplantation of a cell sheet composed of ADSC versus fibroblasts, demonstrating the superior effect of the mesenchymal progenitors (Miyahara et al. 2006). Recently, autologous ADSC were co-transplanted with allogenic human embryonic stem cell-derived CD15+cardiac progenitors in a monkey model of MI, providing proof of the safety of the procedure, although functionality was not analyzed due to the small number of treated animals (Bel et al. 2010).

Stem cell differentiation, though shown not to be necessary for a therapy to induce significant benefits (Mazo et al. 2010a, Mazo et al. 2011), is one of the most desired goals in the field. If transplanted cells are not able to replace ischemia-damaged cells, the treatment may limit infarct expansion, but the damaged tissue will not be substituted, leaving the organ chronically scarred. As a consequence, some reports have aimed at increasing MSC differentiation in vitro prior to its transplantation. Hahn and coworkers incubated cells in a mixture of fibroblast growth factor 2, IGF1 and bone morphogenetic protein 2 (all of them well known promoters of cardiac differentiation) (Hahn et al. 2008). In culture, pre-treated cells, though not differentiated, were able to protect neighboring CM from hypoxia via enhanced gap junctional communication, which was in vivo translated to a significant benefit in cardiac function even in comparison with untreated cells. Cho et al. lentivirally transduced BM-MSC to overexpress GSK3β under the rationale of increasing cell differentiation to CM. When transplanted in a mouse model of MI, the treatment showed an enhanced efficacy as well as an improved cell differentiation capacity (Cho et al. 2011). Tissue architecture and mechanical properties have also been revealed as principal mediators of cell differentiation (Engler et al. 2006, 2008). Although few, some published reports are starting to investigate the effect of substrate geometry and rigidity on MSC differentiation, with striking results (Engler et al. 2006; Forte et al. 2008).

Finally, the immune modulation capacity of mesenchymal progenitors (reviewed in Le Blanc 2006), which in theory provides the means for the allogenic use of MSCs as an off-the-shelf product (expanded prior to the onset of the ischemia and applicable on demand), needs to be carefully revised. Two reports have compared the effects of allogenic versus syngenic injection of BM-MSC in a rat model of MI, with conflicting results. Imanishi et al. (Imanishi et al. 2008) demonstrated that both autologous and allogenic cells improved cardiac function 4 weeks after transplantation, remained in the damaged tissues and did not stimulate rejection. Huang and coworkers, on the other hand (Huang et al. 2010b), followed animals for up to 6 months. Syngenic cells stimulated cardiac recovery, whereas the effect of the allogenic treatment was transitory (significant 3 months post-injection but not at 6) and BM-MSC disappeared earlier than their syngenic counterpart. However, this difference can be attributed to methodological discrepancies regarding the time of transplantation (acute versus chronic respectively) or the follow-up (1 versus 6 months). Importantly, results from clinically relevant large animal models of MI in which allogenic cells have been employed have revealed either positive (Schuleri et al. 2008; Amado et al. 2006) or no functional outcome (Hashemi et al. 2008). On the other hand, when autologous ADSCs or BM-MSCs were used (Dubois et al. 2010; Cai et al. 2009a; Rigol et al. 2010; Zhou et al. 2009), reports have shown a robust and consistent functional recovery after cell transplantation. Thus, although consistently demonstrated in vitro, in vivo immune modulation by MSC must be accurately assayed.

Conclusion

Mesenchymal cells have received substantial interest in recent years as a potential therapy for MI. Endowed with certain differentiation ability, potent paracrine activity and capacity for immune modulation, MSC have been one of the major candidates for reaching the bedside. Nevertheless, the information already gathered from pre- and clinical studies has significantly shifted this picture to one where immune modulation in the setting of MI must be carefully revised and in vitro transdifferentiation has not been translated into any significant contribution to vascular or cardiac lineages. In spite of these shortcomings, the paracrine action of transplanted cells is sufficiently potent to be capable of inducing an important benefit in terms not only of cardiac contractility but also histology. The next challenge faced by both scientists and clinicians is to better understand the pleiotropic capacities of MSC before further improvements in therapy can be achieved.

Source of Funding This work was supported by grants to FP (FP7 HEALTH-F2-2009-222995 INELPY, Instituto de Salud Carlos III (ISCIII-RETIC RD06/0014 (FIS)) and Ministerio de Ciencia e Innovación Programa de Internacionalización (CARDIOBIO) and BP (Instituto de Salud Carlos III (PI10/01621 and CP09/00333 (FIS))).

References

Aggarwal S, Pittenger MF (2005) Human mesenchymal stem cells modulate allogeneic immune cell responses. Blood 105(4):1815–1822. doi:10.1182/blood-2004-04-1559

Amado LC, Saliaris AP, Schuleri KH, St John M, Xie JS, Cattaneo S, Durand DJ, Fitton T, Kuang JQ, Stewart G, Lehrke S, Baumgartner WW, Martin BJ, Heldman AW, Hare JM (2005) Cardiac repair with intramyocardial injection of allogeneic mesenchymal stem cells after myocardial infarction. Proc Natl Acad Sci U. S. A. 102(32):11474–11479. doi:10.1073/pnas.0504388102

Amado LC, Schuleri KH, Saliaris AP, Boyle AJ, Helm R, Oskouei B, Centola M, Eneboe V, Young R, Lima JA, Lardo AC, Heldman AW, Hare JM (2006) Multimodality noninvasive imaging demonstrates in vivo cardiac regeneration after mesenchymal stem cell therapy. J Am Coll Cardiol 48(10):2116–2124. doi:10.1016/j.jacc.2006.06.073

Arminan A, Gandia C, Garcia-Verdugo JM, Lledo E, Mullor JL, Montero JA, Sepulveda P (2010) Cardiac transcription factors driven lineage-specification of adult stem cells. J Cardiovasc Transl Res 3(1):61–65. doi:10.1007/s12265-009-9144-3

Bai X, Yan Y, Song YH, Seidensticker M, Rabinovich B, Metzele R, Bankson JA, Vykoukal D, Alt E (2010) Both cultured and freshly isolated adipose tissue-derived stem cells enhance cardiac function after acute myocardial infarction. Eur Heart J 31(4):489–501. doi:10.1093/eurheartj/ehp568

Bai X, Yan Y, Coleman M, Wu G, Rabinovich B, Seidensticker M, Alt E (2011) Tracking long-term survival of intramyocardially delivered human adipose tissue-derived stem cells using bioluminescence imaging. Mol Imaging Biol 13(4):633–645. doi:10.1007/s11307-010-0392-z

Bel A, Planat-Bernard V, Saito A, Bonnevie L, Bellamy V, Sabbah L, Bellabas L, Brinon B, Vanneaux V, Pradeau P, Peyrard S, Larghero J, Pouly J, Binder P, Garcia S, Shimizu T, Sawa Y, Okano T, Bruneval P, Desnos M, Hagege AA, Casteilla L, Puceat M, Menasche P (2010) Composite cell sheets: a further step toward safe and effective myocardial regeneration by cardiac progenitors derived from embryonic stem cells. Circulation 122(11 Suppl):118–123. doi:10.1161/CIRCULATIONAHA.109.927293

Bianco P, Robey PG, Simmons PJ (2008) Mesenchymal stem cells: revisiting history, concepts, and assays. Cell Stem Cell 2(4):313–319. doi:10.1016/j.stem.2008.03.002

Cai L, Johnstone BH, Cook TG, Tan J, Fishbein MC, Chen PS, March KL (2009a) IFATS collection: human adipose tissue-derived stem cells induce angiogenesis and nerve sprouting following myocardial infarction, in conjunction with potent preservation of cardiac function. Stem Cells 27(1):230–237. doi:10.1634/stemcells.2008-0273

Cai X, Lin Y, Friedrich CC, Neville C, Pomerantseva I, Sundback CA, Zhang Z, Vacanti JP, Hauschka PV, Grottkau BE (2009b) Bone marrow derived pluripotent cells are pericytes which contribute to vascularization. Stem Cell Rev 5(4):437–445. doi:10.1007/s12015-009-9097-6

Caplan AI, Dennis JE (2006) Mesenchymal stem cells as trophic mediators. J Cell Biochem 98(5):1076–1084. doi:10.1002/jcb.20886

Carr CA, Stuckey DJ, Tatton L, Tyler DJ, Hale SJ, Sweeney D, Schneider JE, Martin-Rendon E, Radda GK, Harding SE, Watt SM, Clarke K (2008) Bone marrow-derived stromal cells home to and remain in the infarcted rat heart but fail to improve function: an in vivo cine-MRI study. Am J Physiol Heart Circ Physiol 295(2):533–542. doi:10.1152/ajpheart.00094.2008

Chang SA, Lee EJ, Kang HJ, Zhang SY, Kim JH, Li L, Youn SW, Lee CS, Kim KH, Won JY, Sohn JW, Park KW, Cho HJ, Yang SE, Oh WI, Yang YS, Ho WK, Park YB, Kim HS (2008) Impact of myocardial infarct proteins and oscillating pressure on the differentiation of mesenchymal stem cells: effect of acute myocardial infarction on stem cell differentiation. Stem Cells 26(7):1901–1912. doi:10.1634/stemcells.2007-0708

Chen CH, Wei HJ, Lin WW, Chiu I, Hwang SM, Wang CC, Lee WY, Chang Y, Sung HW (2008a) Porous tissue grafts sandwiched with multilayered mesenchymal stromal cell sheets induce tissue regeneration for cardiac repair. Cardiovasc Res 80(1):88–95. doi:10.1093/cvr/cvn149

Chen L, Tredget EE, Wu PY, Wu Y (2008b) Paracrine factors of mesenchymal stem cells recruit macrophages and endothelial lineage cells and enhance wound healing. PLoS One 3(4):e1886. doi:10.1371/journal.pone.0001886

Cheng Z, Liu X, Ou L, Zhou X, Liu Y, Jia X, Zhang J, Li Y, Kong D (2008a) Mobilization of mesenchymal stem cells by granulocyte colony-stimulating factor in rats with acute myocardial infarction. Cardiovasc Drugs Ther 22(5):363–371. doi:10.1007/s10557-008-6110-2

Cheng Z, Ou L, Zhou X, Li F, Jia X, Zhang Y, Liu X, Li Y, Ward CA, Melo LG, Kong D (2008b) Targeted migration of mesenchymal stem cells modified with CXCR4 gene to infarcted myocardium improves cardiac performance. Mol Ther 16(3):571–579. doi:10.1038/sj.mt.6300374

Cho J, Zhai P, Maejima Y, Sadoshima J (2011) Myocardial injection with GSK-3beta-overexpressing bone marrow-derived mesenchymal stem cells attenuates cardiac dysfunction after myocardial infarction. Circ Res 108(4):478–489. doi:10.1161/CIRCRESAHA.110.229658

Cleutjens JP, Kandala JC, Guarda E, Guntaka RV, Weber KT (1995) Regulation of collagen degradation in the rat myocardium after infarction. J Mol Cell Cardiol 27(6):1281–1292

Crop MJ, Baan CC, Korevaar SS, Ijzermans JN, Pescatori M, Stubbs AP, van Ijcken WF, Dahlke MH, Eggenhofer E, Weimar W, Hoogduijn MJ (2010) Inflammatory conditions affect gene expression and function of human adipose tissue-derived mesenchymal stem cells. Clin Exp Immunol 162(3):474–486. doi:10.1111/j.1365-2249.2010.04256.x

de Macedo Braga LM, Lacchini S, Schaan BD, Rodrigues B, Rosa K, De Angelis K, Borges LF, Irigoyen MC, Nardi NB (2008) In situ delivery of bone marrow cells and mesenchymal stem cells improves cardiovascular function in hypertensive rats submitted to myocardial infarction. J Biomed Sci 15(3):365–374. doi:10.1007/s11373-008-9237-z

De Ugarte DA, Alfonso Z, Zuk PA, Elbarbary A, Zhu M, Ashjian P, Benhaim P, Hedrick MH, Fraser JK (2003) Differential expression of stem cell mobilization-associated molecules on multi-lineage cells from adipose tissue and bone marrow. Immunol Lett 89(2-3):267–270

Dimmeler S, Leri A (2008) Aging and disease as modifiers of efficacy of cell therapy. Circ Res 102(11):1319–1330. doi:10.1161/CIRCRESAHA.108.175943

Dixon JA, Gorman RC, Stroud RE, Bouges S, Hirotsugu H, Gorman JH 3rd, Martens TP, Itescu S, Schuster MD, Plappert T, St John-SuttonMG, Spinale FG (2009) Mesenchymal cell transplantation and myocardial remodeling after myocardial infarction. Circulation 120(11 Suppl):220–229. doi:10.1161/CIRCULATIONAHA.108.842302

Dominici M, Le Blanc K, Mueller I, Slaper-Cortenbach I, Marini F, Krause D, Deans R, Keating A, Prockop D, Horwitz E (2006) Minimal criteria for defining multipotent mesenchymal stromal cells. The International Society for Cellular Therapy position statement. Cytotherapy 8(4):315–317. doi:10.1080/14653240600855905

Dubois C, Liu X, Claus P, Marsboom G, Pokreisz P, Vandenwijngaert S, Depelteau H, Streb W, Chaothawee L, Maes F, Gheysens O, Debyser Z, Gillijns H, Pellens M, Vandendriessche T, Chuah M, Collen D, Verbeken E, Belmans A, Van de Werf F, Bogaert J, Janssens S (2010) Differential effects of progenitor cell populations on left ventricular remodeling and myocardial neovascularization after myocardial infarction. J Am Coll Cardiol 55(20):2232–2243. doi:10.1016/j.jacc.2009.10.081

Engler AJ, Sen S, Sweeney HL, Discher DE (2006) Matrix elasticity directs stem cell lineage specification. Cell 126(4):677–689. doi:10.1016/j.cell.2006.06.044

Engler AJ, Carag-Krieger C, Johnson CP, Raab M, Tang HY, Speicher DW, Sanger JW, Sanger JM, Discher DE (2008) Embryonic cardiomyocytes beat best on a matrix with heart-like elasticity: scar-like rigidity inhibits beating. J Cell Sci 121(Pt 22):3794–3802. doi:10.1242/jcs.029678

Fedak PW (2008) Paracrine effects of cell transplantation: modifying ventricular remodeling in the failing heart. Semin Thorac Cardiovasc Surg 20(2):87–93. doi:10.1053/j.semtcvs.2008.04.001

Forte G, Carotenuto F, Pagliari F, Pagliari S, Cossa P, Fiaccavento R, Ahluwalia A, Vozzi G, Vinci B, Serafino A, Rinaldi A, Traversa E, Carosella L, Minieri M, Di Nardo P (2008) Criticality of the biological and physical stimuli array inducing resident cardiac stem cell determination. Stem Cells 26(8):2093–2103. doi:10.1634/stemcells.2008-0061

Friedenstein AJ, Chailakhyan RK, Latsinik NV, Panasyuk AF, Keiliss-Borok IV (1974a) Stromal cells responsible for transferring the microenvironment of the hemopoietic tissues. Cloning in vitro and retransplantation in vivo. Transplantation 17(4):331–340

Friedenstein AJ, Deriglasova UF, Kulagina NN, Panasuk AF, Rudakowa SF, Luria EA, Ruadkow IA (1974b) Precursors for fibroblasts in different populations of hematopoietic cells as detected by the in vitro colony assay method. Exp Hematol 2(2):83–92

Gaebel R, Furlani D, Sorg H, Polchow B, Frank J, Bieback K, Wang W, Klopsch C, Ong LL, Li W, Ma N, Steinhoff G (2011) Cell origin of human mesenchymal stem cells determines a different healing performance in cardiac regeneration. PLoS One 6(2):e15652. doi:10.1371/journal.pone.0015652

Gandia C, Arminan A, Garcia-Verdugo JM, Lledo E, Ruiz A, Minana MD, Sanchez-Torrijos J, Paya R, Mirabet V, Carbonell-Uberos F, Llop M, Montero JA, Sepulveda P (2008) Human dental pulp stem cells improve left ventricular function, induce angiogenesis, and reduce infarct size in rats with acute myocardial infarction. Stem Cells 26(3):638–645. doi:10.1634/stemcells.2007-0484

Gimble JM, Katz AJ, Bunnell BA (2007) Adipose-derived stem cells for regenerative medicine. Circ Res 100(9):1249–1260. doi:10.1161/01.RES.0000265074.83288.09

Gnecchi M, He H, Liang OD, Melo LG, Morello F, Mu H, Noiseux N, Zhang L, Pratt RE, Ingwall JS, Dzau VJ (2005) Paracrine action accounts for marked protection of ischemic heart by Akt-modified mesenchymal stem cells. Nat Med 11(4):367–368. doi:10.1038/nm0405-367

Gnecchi M, He H, Noiseux N, Liang OD, Zhang L, Morello F, Mu H, Melo LG, Pratt RE, Ingwall JS, Dzau VJ (2006) Evidence supporting paracrine hypothesis for Akt-modified mesenchymal stem cell-mediated cardiac protection and functional improvement. FASEB J 20(6):661–669. doi:10.1096/fj.05-5211com

Gnecchi M, He H, Melo LG, Noiseaux N, Morello F, de Boer RA, Zhang L, Pratt RE, Dzau VJ, Ingwall JS (2009) Early beneficial effects of bone marrow-derived mesenchymal stem cells overexpressing Akt on cardiac metabolism after myocardial infarction. Stem Cells 27(4):971–979. doi:10.1002/stem.12

Hahn JY, Cho HJ, Kang HJ, Kim TS, Kim MH, Chung JH, Bae JW, Oh BH, Park YB, Kim HS (2008) Pre-treatment of mesenchymal stem cells with a combination of growth factors enhances gap junction formation, cytoprotective effect on cardiomyocytes, and therapeutic efficacy for myocardial infarction. J Am Coll Cardiol 51(9):933–943. doi:10.1016/j.jacc.2007.11.040

Haider H, Ashraf M (2008) Strategies to promote donor cell survival: combining preconditioning approach with stem cell transplantation. J Mol Cell Cardiol 45(4):554–566. doi:10.1016/j.yjmcc.2008.05.004

Hale SL, Dai W, Dow JS, Kloner RA (2008) Mesenchymal stem cell administration at coronary artery reperfusion in the rat by two delivery routes: a quantitative assessment. Life Sci 83(13-14):511–515. doi:10.1016/j.lfs.2008.07.020

Hashemi SM, Ghods S, Kolodgie FD, Parcham-Azad K, Keane M, Hamamdzic D, Young R, Rippy MK, Virmani R, Litt H, Wilensky RL (2008) A placebo controlled, dose-ranging, safety study of allogenic mesenchymal stem cells injected by endomyocardial delivery after an acute myocardial infarction. Eur Heart J 29(2):251–259. doi:10.1093/eurheartj/ehm559

Hida N, Nishiyama N, Miyoshi S, Kira S, Segawa K, Uyama T, Mori T, Miyado K, Ikegami Y, Cui C, Kiyono T, Kyo S, Shimizu T, Okano T, Sakamoto M, Ogawa S, Umezawa A (2008) Novel cardiac precursor-like cells from human menstrual blood-derived mesenchymal cells. Stem Cells 26(7):1695–1704. doi:10.1634/stemcells.2007-0826

Horwitz EM, Le Blanc K, Dominici M, Mueller I, Slaper-Cortenbach I, Marini FC, Deans RJ, Krause DS, Keating A, International Society for Cellular T (2005) Clarification of the nomenclature for MSC: The International Society for Cellular Therapy position statement. Cyto-Therapy 7(5):393–395. doi:10.1080/14653240500319234

Huang J, Zhang Z, Guo J, Ni A, Deb A, Zhang L, Mirotsou M, Pratt RE, Dzau VJ (2010a) Genetic modification of mesenchymal stem cells overexpressing CCR1 increases cell viability, migration, engraftment, and capillary density in the injured myocardium. Circ Res 106(11):1753–1762. doi:10.1161/CIRCRESAHA.109.196030

Huang XP, Sun Z, Miyagi Y, McDonald Kinkaid H, Zhang L, Weisel RD, Li RK (2010b) Differentiation of allogeneic mesenchymal stem cells induces immunogenicity and limits their long-term benefits for myocardial repair. Circulation 122(23):2419–2429. doi:10.1161/CIRCULATIONAHA.110.955971

Ii M, Horii M, Yokoyama A, Shoji T, Mifune Y, Kawamoto A, Asahi M, Asahara T (2011) Synergistic effect of adipose-derived stem cell therapy and bone marrow progenitor recruitment in ischemic heart. Lab Invest 91(4):539–552. doi:10.1038/labinvest.2010.191

Imanishi Y, Saito A, Komoda H, Kitagawa-Sakakida S, Miyagawa S, Kondoh H, Ichikawa H, Sawa Y (2008) Allogenic mesenchymal stem cell transplantation has a therapeutic effect in acute myocardial infarction in rats. J Mol Cell Cardiol 44(4):662–671. doi:10.1016/j.yjmcc.2007.11.001

Jain R, Ellika SK, Scarpace L, Schultz LR, Rock JP, Gutierrez J, Patel SC, Ewing J, Mikkelsen T (2008) Quantitative estimation of permeability surface-area product in astroglial brain tumors using perfusion CT and correlation with histopathologic grade. AJNR Am J Neuroradiol 29(4):694–700. doi:10.3174/ajnr.A0899

Jin J, Jeong SI, Shin YM, Lim KS, Shin H, Lee YM, Koh HC, Kim KS (2009) Transplantation of mesenchymal stem cells within a poly(lactide-co-epsilon-caprolactone) scaffold improves cardiac function in a rat myocardial infarction model. Eur J Heart Fail 11(2):147–153. doi:10.1093/eurjhf/hfn017

Jujo K, Ii M, Losordo DW (2008) Endothelial progenitor cells in neovascularization of infarcted myocardium. J Mol Cell Cardiol 45(4):530–544. doi:10.1016/j.yjmcc.2008.08.003

Kern S, Eichler H, Stoeve J, Kluter H, Bieback K (2006) Comparative analysis of mesenchymal stem cells from bone marrow, umbilical cord blood, or adipose tissue. Stem Cells 24(5):1294–1301. doi:10.1634/stemcells.2005-0342

Kilroy GE, Foster SJ, Wu X, Ruiz J, Sherwood S, Heifetz A, Ludlow JW, Stricker DM, Potiny S, Green P, Halvorsen YD, Cheatham B, Storms RW, Gimble JM (2007) Cytokine profile of human adipose-derived stem cells: expression of angiogenic, hematopoietic, and pro-inflammatory factors. J Cell Physiol 212(3):702–709. doi:10.1002/jcp.21068

Kim YM, Jeon ES, Kim MR, Jho SK, Ryu SW, Kim JH (2008) Angiotensin II-induced differentiation of adipose tissue-derived mesenchymal stem cells to smooth muscle-like cells. Int J Biochem Cell Biol 40(11):2482–2491. doi:10.1016/j.biocel.2008.04.016

Kondo T, Hayashi M, Takeshita K, Numaguchi Y, Kobayashi K, Iino S, Inden Y, Murohara T (2004) Smoking cessation rapidly increases circulating progenitor cells in peripheral blood in chronic smokers. Arterioscler Thromb Vasc Biol 24(8):1442–1447. doi:10.1161/01.ATV.0000135655.52088.c5

Le Blanc K (2006) Mesenchymal stromal cells: tissue repair and immune modulation. Cytotherapy 8(6):559–561. doi:10.1080/14653240601045399

Le Blanc K, Frassoni F, Ball L, Locatelli F, Roelofs H, Lewis I, Lanino E, Sundberg B, Bernardo ME, Remberger M, Dini G, Egeler RM, Bacigalupo A, Fibbe W, Ringden O, Developmental Committee of the European Group for B, Marrow T (2008) Mesenchymal stem cells for treatment of steroid-resistant, severe, acute graft-versus-host disease: a phase II study. Lancet 371(9624):1579–1586. doi:10.1016/S0140-6736(08)60690-X

Lee BC, Hsu HC, Tseng WY, Chen CY, Lin HJ, Ho YL, Su MJ, Chen MF (2009) Cell therapy generates a favourable chemokine gradient for stem cell recruitment into the infarcted heart in rabbits. Eur J Heart Fail 11(3):238–245. doi:10.1093/eurjhf/hfn035

Lee OK, Kuo TK, Chen WM, Lee KD, Hsieh SL, Chen TH (2004) Isolation of multipotent mesenchymal stem cells from umbilical cord blood. Blood 103(5):1669–1675. doi:10.1182/blood-2003-05-1670

Leobon B, Roncalli J, Joffre C, Mazo M, Boisson M, Barreau C, Calise D, Arnaud E, Andre M, Puceat M, Penicaud L, Prosper F, Planat-Benard V, Casteilla L (2009) Adipose-derived cardiomyogenic cells: in vitro expansion and functional improvement in a mouse model of myocardial infarction. Cardiovasc Res 83(4):757–767. doi:10.1093/cvr/cvp167

Li B, Zeng Q, Wang H, Shao S, Mao X, Zhang F, Li S, Guo Z (2007) Adipose tissue stromal cells transplantation in rats of acute myocardial infarction. Coron Artery Dis 18(3):221–227. doi:10.1097/MCA.0b013e32801235da

Li H, Malhotra D, Yeh CC, Tu R, Zhu BQ, Birger N, Wisneski A, Cha J, Karliner JS, Mann MJ (2009a) Myocardial survival signaling in response to stem cell transplantation. J Am Coll Surg 208(4):607–613. doi:10.1016/j.jamcollsurg.2008.12.016

Li L, Zhang S, Zhang Y, Yu B, Xu Y, Guan Z (2009b) Paracrine action mediate the antifibrotic effect of transplanted mesenchymal stem cells in a rat model of global heart failure. Mol Biol Rep 36(4):725–731. doi:10.1007/s11033-008-9235-2

Liang H, Hou H, Yi W, Yang G, Gu C, Lau WB, Gao E, Ma X, Lu Z, Wei X, Pei J, Yi D (2011) Increased expression of pigment epithelium-derived factor in aged mesenchymal stem cells impairs their therapeutic efficacy for attenuating myocardial infarction injury. Eur Heart J 34(22):1681–1690. doi:10.1093/eurheartj/ehr131

Liu JW, Dunoyer-Geindre S, Serre-Beinier V, Mai G, Lambert JF, Fish RJ, Pernod G, Buehler L, Bounameaux H, Kruithof EK (2007) Characterization of endothelial-like cells derived from human mesenchymal stem cells. J Thromb Haemost 5(4):826–834. doi:10.1111/j.1538-7836.2007.02381.x

Liu XH, Bai CG, Xu ZY, Huang SD, Yuan Y, Gong DJ, Zhang JR (2008) Therapeutic potential of angiogenin modified mesenchymal stem cells: angiogenin improves mesenchymal stem cells survival under hypoxia and enhances vasculogenesis in myocardial infarction. Microvasc Res 76(1):23–30. doi:10.1016/j.mvr.2008.02.005

Lozito TP, Taboas JM, Kuo CK, Tuan RS (2009) Mesenchymal stem cell modification of endothelial matrix regulates their vascular differentiation. J Cell Biochem 107(4):706–713. doi:10.1002/jcb.22166

Luan X, Li G, Wang G, Wang F, Lin Y (2013) Human placenta-derived mesenchymal stem cells suppress T cell proliferation and support the culture expansion of cord blood CD34(+) cells: a comparison with human bone marrow-derived mesenchymal stem cells. Tissue Cell 45(1):32–38. doi:10.1016/j.tice.2012.09.002

Ma J, Ge J, Zhang S, Sun A, Shen J, Chen L, Wang K, Zou Y (2005) Time course of myocardial stromal cell-derived factor 1 expression and beneficial effects of intravenously administered bone marrow stem cells in rats with experimental myocardial infarction. Basic Res Cardiol 100(3):217–223. doi:10.1007/s00395-005-0521-z

Madonna R, Renna FV, Cellini C, Cotellese R, Picardi N, Francomano F, Innocenti P, De Caterina R (2011) Age-dependent impairment of number and angiogenic potential of adipose tissue-derived progenitor cells. Eur J Clin Invest 41(2):126–133. doi:10.1111/j.1365-2362.2010.02384.x

Mann DL (1999) Mechanisms and models in heart failure: a combinatorial approach. Circulation 100(9):999–1008

Mazo M, Planat-Benard V, Abizanda G, Pelacho B, Leobon B, Gavira JJ, Penuelas I, Cemborain A, Penicaud L, Laharrague P, Joffre C, Boisson M, Ecay M, Collantes M, Barba J, Casteilla L, Prosper F (2008) Transplantation of adipose derived stromal cells is associated with functional improvement in a rat model of chronic myocardial infarction. Eur J Heart Fail 10(5):454–462. doi:10.1016/j.ejheart.2008.03.017

Mazo M, Gavira JJ, Abizanda G, Moreno C, Ecay M, Soriano M, Aranda P, Collantes M, Alegria E, Merino J, Penuelas I, Garcia Verdugo JM, Pelacho B, Prosper F (2010a) Transplantation of mesenchymal stem cells exerts a greater long-term effect than bone marrow mononuclear cells in a chronic myocardial infarction model in rat. Cell Transplant 19(3):313–328. doi:10.3727/096368909X480323

Mazo M, Pelacho B, Prosper F (2010b) Stem cell therapy for chronic myocardial infarction. J Cardiovasc Transl Res 3(2):79–88. doi:10.1007/s12265-009-9159-9

Mazo M, Gavira JJ, Pelacho B, Prosper F (2011) Adipose-derived stem cells for myocardial infarction. J Cardiovasc Transl Res 4(2):145–153. doi:10.1007/s12265-010-9246-y

McIntosh K, Zvonic S, Garrett S, Mitchell JB, Floyd ZE, Hammill L, Kloster A, Di Halvorsen Y, Ting JP, Storms RW, Goh B, Kilroy G, Wu X, Gimble JM (2006) The immunogenicity of human adipose-derived cells: temporal changes in vitro. Stem Cells 24(5):1246–1253. doi:10.1634/stemcells.2005-0235

Menasche P (2011) Cardiac cell therapy: lessons from clinical trials. J Mol Cell Cardiol 50(2):258–265. doi:10.1016/j.yjmcc.2010.06.010

Mias C, Lairez O, Trouche E, Roncalli J, Calise D, Seguelas MH, Ordener C, Piercecchi-Marti MD, Auge N, Salvayre AN, Bourin P, Parini A, Cussac D (2009) Mesenchymal stem cells promote matrix metalloproteinase secretion by cardiac fibroblasts and reduce cardiac ventricular fibrosis after myocardial infarction. Stem Cells 27(11):2734–2743. doi:10.1002/stem.169

Mitchell JB, McIntosh K, Zvonic S, Garrett S, Floyd ZE, Kloster A, Di Halvorsen Y, Storms RW, Goh B, Kilroy G, Wu X, Gimble JM (2006) Immunophenotype of human adipose-derived cells: temporal changes in stromal-associated and stem cell-associated markers. Stem Cells 24(2):376–385. doi:10.1634/stemcells.2005-0234

Miyahara Y, Nagaya N, Kataoka M, Yanagawa B, Tanaka K, Hao H, Ishino K, Ishida H, Shimizu T, Kangawa K, Sano S, Okano T, Kitamura S, Mori H (2006) Monolayered mesenchymal stem cells repair scarred myocardium after myocardial infarction. Nat Med 12(4):459–465. doi:10.1038/nm1391

Nahrendorf M, Swirski FK, Aikawa E, Stangenberg L, Wurdinger T, Figueiredo JL, Libby P, Weissleder R, Pittet MJ (2007) The healing myocardium sequentially mobilizes two monocyte subsets with divergent and complementary functions. J Exp Med 204(12):3037–3047. doi:10.1084/jem.20070885

Nian M, Lee P, Khaper N, Liu P (2004) Inflammatory cytokines and postmyocardial infarction remodeling. Circ Res 94(12):1543–1553. doi:10.1161/01.RES.0000130526.20854.fa

Ohnishi S, Sumiyoshi H, Kitamura S, Nagaya N (2007) Mesenchymal stem cells attenuate cardiac fibroblast proliferation and collagen synthesis through paracrine actions. FEBS Lett 581(21):3961–3966. doi:10.1016/j.febslet.2007.07.028

Organization WH (2008) World Health Statistics 2008.

Oskowitz A, McFerrin H, Gutschow M, Carter ML, Pochampally R (2011) Serum-deprived human multipotent mesenchymal stromal cells (MSCs) are highly angiogenic. Stem Cell Res 6(3):215–225. doi:10.1016/j.scr.2011.01.004

Penn MS (2009) Importance of the SDF-1:CXCR4 axis in myocardial repair. Circ Res 104(10):1133–1135. doi:10.1161/CIRCRESAHA.109.198929

Pittenger MF, Mackay AM, Beck SC, Jaiswal RK, Douglas R, Mosca JD, Moorman MA, Simonetti DW, Craig S, Marshak DR (1999) Multilineage potential of adult human mesenchymal stem cells. Science 284(5411):143–147

Planat-Benard V, Silvestre JS, Cousin B, Andre M, Nibbelink M, Tamarat R, Clergue M, Manneville C, Saillan-Barreau C, Duriez M, Tedgui A, Levy B, Penicaud L, Casteilla L (2004) Plasticity of human adipose lineage cells toward endothelial cells: physiological and therapeutic perspectives. Circulation 109(5):656–663. doi:10.1161/01.CIR.0000114522.38265.61

Puymirat E, Geha R, Tomescot A, Bellamy V, Larghero J, Trinquart L, Bruneval P, Desnos M, Hagege A, Puceat M, Menasche P (2009) Can mesenchymal stem cells induce tolerance to cotransplanted human embryonic stem cells? Mol Ther 17(1):176–182. doi:10.1038/mt.2008.208

Rehman J, Traktuev D, Li J, Merfeld-Clauss S, Temm-Grove CJ, Bovenkerk JE, Pell CL, Johnstone BH, Considine RV, March KL (2004) Secretion of angiogenic and antiapoptotic factors by human adipose stromal cells. Circulation 109(10):1292–1298. doi:10.1161/01.CIR.0000121425.42966.F1

Reinecke H, Minami E, Zhu WZ, Laflamme MA (2008) Cardiogenic differentiation and transdifferentiation of progenitor cells. Circ Res 103(10):1058–1071. doi:10.1161/CIRCRESAHA.108.180588

Rigol M, Solanes N, Farre J, Roura S, Roque M, Berruezo A, Bellera N, Novensa L, Tamborero D, Prat-Vidal C, Huzman MA, Batlle M, Hoefsloot M, Sitges M, Ramirez J, Dantas AP, Merino A, Sanz G, Brugada J, Bayes-Genis A, Heras M (2010) Effects of adipose tissue-derived stem cell therapy after myocardial infarction: impact of the route of administration. J Card Fail 16(4):357–366. doi:10.1016/j.cardfail.2009.12.006

Robey TE, Saiget MK, Reinecke H, Murry CE (2008) Systems approaches to preventing transplanted cell death in cardiac repair. J Mol Cell Cardiol 45(4):567–581. doi:10.1016/j.yjmcc.2008.03.009

Rogers TB, Pati S, Gaa S, Riley D, Khakoo AY, Patel S, Wardlow RD 2nd, Frederick CA, Hall G, He LP, Lederer WJ (2011) Mesenchymal stem cells stimulate protective genetic reprogramming of injured cardiac ventricular myocytes. J Mol Cell Cardiol 50(2):346–356. doi:10.1016/j.yjmcc.2010.09.001

Sacchetti B, Funari A, Michienzi S, Di Cesare S, Piersanti S, Saggio I, Tagliafico E, Ferrari S, Robey PG, Riminucci M, Bianco P (2007) Self-renewing osteoprogenitors in bone marrow sinusoids can organize a hematopoietic microenvironment. Cell 131(2):324–336. doi:10.1016/j.cell.2007.08.025

Schuleri KH, Amado LC, Boyle AJ, Centola M, Saliaris AP, Gutman MR, Hatzistergos KE, Oskouei BN, Zimmet JM, Young RG, Heldman AW, Lardo AC, Hare JM (2008) Early improvement in cardiac tissue perfusion due to mesenchymal stem cells. Am J Physiol Heart Circ Physiol 294(5):2002–2011. doi:10.1152/ajpheart.00762.2007

Sengenes C, Miranville A, Maumus M, de Barros S, Busse R, Bouloumie A (2007) Chemotaxis and differentiation of human adipose tissue CD34+/CD31- progenitor cells: role of stromal derived factor-1 released by adipose tissue capillary endothelial cells. Stem Cells 25(9): 2269–2276. doi:10.1634/stemcells.2007-0180

Song H, Cha MJ, Song BW, Kim IK, Chang W, Lim S, Choi EJ, Ham O, Lee SY, Chung N, Jang Y, Hwang KC (2010) Reactive oxygen species inhibit adhesion of mesenchymal stem cells implanted into ischemic myocardium via interference of focal adhesion complex. Stem Cells 28(3):555–563. doi:10.1002/stem.302

Song YH, Gehmert S, Sadat S, Pinkernell K, Bai X, Matthias N, Alt E (2007) VEGF is critical for spontaneous differentiation of stem cells into cardiomyocytes. Biochem Biophys Res Commun 354(4):999–1003. doi:10.1016/j.bbrc.2007.01.095

Sun N, Panetta NJ, Gupta DM, Wilson KD, Lee A, Jia F, Hu S, Cherry AM, Robbins RC, Longaker MT, Wu JC (2009) Feeder-free derivation of induced pluripotent stem cells from adult human adipose stem cells. Proc Natl Acad Sci U. S. A. 106(37):15720–15725. doi:10.1073/pnas.0908450106

Sun Y, Kiani MF, Postlethwaite AE, Weber KT (2002) Infarct scar as living tissue. Basic Res Cardiol 97(5):343–347. doi:10.1007/s00395-002-0365-8

Suzuki G, Iyer V, Lee TC, Canty JM Jr (2011) Autologous mesenchymal stem cells mobilize cKit+ and CD133+ bone marrow progenitor cells and improve regional function in hibernating myocardium. Circ Res 109(9):1044–1054. doi:10.1161/CIRCRESAHA.111.245969

Takahashi K, Yamanaka S (2006) Induction of pluripotent stem cells from mouse embryonic and adult fibroblast cultures by defined factors. Cell 126(4):663–676. doi:10.1016/j.cell.2006.07.024

Tan MY, Zhi W, Wei RQ, Huang YC, Zhou KP, Tan B, Deng L, Luo JC, Li XQ, Xie HQ, Yang ZM (2009) Repair of infarcted myocardium using mesenchymal stem cell seeded small intestinal submucosa in rabbits. Biomaterials 30(19):3234–3240. doi:10.1016/j.biomaterials.2009.02.013

Tat PA, Sumer H, Jones KL, Upton K, Verma PJ (2010) The efficient generation of induced pluripotent stem (iPS) cells from adult mouse adipose tissue-derived and neural stem cells. Cell Transplant 19(5):525–536. doi:10.3727/096368910X491374

Traktuev DO, Merfeld-Clauss S, Li J, Kolonin M, Arap W, Pasqualini R, Johnstone BH, March KL (2008) A population of multipotent CD34-positive adipose stromal cells share pericyte and mesenchymal surface markers, reside in a periendothelial location, and stabilize endothelial networks. Circ Res 102(1):77–85. doi:10.1161/CIRCRESAHA.107.159475

Traktuev DO, Prater DN, Merfeld-Clauss S, Sanjeevaiah AR, Saadatzadeh MR, Murphy M, Johnstone BH, Ingram DA, March KL (2009) Robust functional vascular network formation in vivo by cooperation of adipose progenitor and endothelial cells. Circ Res 104(12):1410–1420. doi:10.1161/CIRCRESAHA.108.190926

Valina C, Pinkernell K, Song YH, Bai X, Sadat S, Campeau RJ, Le Jemtel TH, Alt E (2007) Intracoronary administration of autologous adipose tissue-derived stem cells improves left ventricular function, perfusion, and remodelling after acute myocardial infarction. Eur Heart J 28(21):2667–2677. doi:10.1093/eurheartj/ehm426

van der Bogt KE, Sheikh AY, Schrepfer S, Hoyt G, Cao F, Ransohoff KJ, Swijnenburg RJ, Pearl J, Lee A, Fischbein M, Contag CH, Robbins RC, Wu JC (2008) Comparison of different adult stem cell types for treatment of myocardial ischemia. Circulation 118(14 Suppl):121–129. doi:10.1161/CIRCULATIONAHA.107.759480

van der Bogt KE, Schrepfer S, Yu J, Sheikh AY, Hoyt G, Govaert JA, Velotta JB, Contag CH, Robbins RC, Wu JC (2009) Comparison of transplantation of adipose tissue- and bone marrow-derived mesenchymal stem cells in the infarcted heart. Transplantation 87(5):642–652. doi:10.1097/TP.0b013e31819609d9 van Dijk A, Niessen HW, Zandieh Doulabi B, Visser FC, van Milligen FJ (2008) Differentiation of human adipose-derived stem cells towards cardiomyocytes is facilitated by laminin. Cell Tissue Res 334(3):457–467. doi:10.1007/s00441-008-0713-6

Vasa M, Fichtlscherer S, Aicher A, Adler K, Urbich C, Martin H, Zeiher AM, Dimmeler S (2001) Number and migratory activity of circulating endothelial progenitor cells inversely correlate with risk factors for coronary artery disease. Circ Res 89(1):E1–7

Virag JI, Murry CE (2003) Myofibroblast and endothelial cell proliferation during murine myocardial infarct repair. Am J Pathol 163(6):2433–2440. doi:10.1016/S0002-9440(10)63598-5

Wagner W, Roderburg C, Wein F, Diehlmann A, Frankhauser M, Schubert R, Eckstein V, Ho AD (2007) Molecular and secretory profiles of human mesenchymal stromal cells and their abilities to maintain primitive hematopoietic progenitors. Stem Cells 25(10):2638–2647. doi:10.1634/stemcells.2007-0280

Wang L, Deng J, Tian W, Xiang B, Yang T, Li G, Wang J, Gruwel M, Kashour T, Rendell J, Glogowski M, Tomanek B, Freed D, Deslauriers R, Arora RC, Tian G (2009a) Adipose-derived stem cells are an effective cell candidate for treatment of heart failure: an MR imaging study of rat hearts. Am J Physiol Heart Circ Physiol 297(3):1020–1031. doi:10.1152/ajpheart.01082.2008

Wang X, Zhao T, Huang W, Wang T, Qian J, Xu M, Kranias EG, Wang Y, Fan GC (2009b) Hsp20-engineered mesenchymal stem cells are resistant to oxidative stress via enhanced activation of Akt and increased secretion of growth factors. Stem Cells 27(12):3021–3031. doi:10.1002/stem.230

White HD, Chew DP (2008) Acute myocardial infarction. The Lancet 372(9638):570–584. doi:10.1016/S0140-6736(08)61237-4

Xu W, Zhang X, Qian H, Zhu W, Sun X, Hu J, Zhou H, Chen Y (2004) Mesenchymal stem cells from adult human bone marrow differentiate into a cardiomyocyte phenotype in vitro. Exp Biol Med (Maywood) 229(7):623–631

Yan X, Lv A, Xing Y, Liu B, Hou J, Huang W, Li Y (2011) Inhibition of p53-p21 pathway promotes the differentiation of rat bone marrow mesenchymal stem cells into cardiomyocytes. Mol Cell Biochem 354(1-2):21–28. doi:10.1007/s11010-011-0801-x

Yang YJ, Qian HY, Huang J, Li JJ, Gao RL, Dou KF, Yang GS, Willerson JT, Geng YJ (2009) Combined therapy with simvastatin and bone marrow-derived mesenchymal stem cells increases benefits in infarcted swine hearts. Arterioscler Thromb Vasc Biol 29(12):2076–2082. doi:10.1161/ATVBAHA.109.189662

Yen BL, Huang HI, Chien CC, Jui HY, Ko BS, Yao M, Shun CT, Yen ML, Lee MC, Chen YC (2005) Isolation of multipotent cells from human term placenta. Stem Cells 23(1):3–9. doi:10.1634/stemcells.2004-0098

Zeng B, Chen H, Zhu C, Ren X, Lin G, Cao F (2008) Effects of combined mesenchymal stem cells and heme oxygenase-1 therapy on cardiac performance. Eur J Cardiothorac Surg 34(4):850–856. doi:10.1016/j.ejcts.2008.05.049

Zhou Y, Wang S, Yu Z, Hoyt RF Jr, Sachdev V, Vincent P, Arai AE, Kwak M, Burkett SS, Horvath KA (2009) Direct injection of autologous mesenchymal stromal cells improves myocardial function. Biochem Biophys Res Commun 390(3):902–907. doi:10.1016/j.bbrc.2009.10.074

Chapter 2
Resident Cardiac Progenitor Cells

Ayelet Itzhaki-Alfia and Jonathan Leor

Abstract Resident cardiac stem/progenitor cells are implicated in cell replacement, repair and maintenance of the myocardium. These cells are already committed to a specific cardiomyogenic pathway. Stimulation of these cells in injured hearts can open new options in cardiovascular regenerative medicine, and advance our approach to patients with end-stage heart failure. This chapter reviews some of the recent discoveries in the field of resident cardiac stem/progenitor cells, focusing first on human cardiac stem cell characteristics and reparative potential, including the concept of myocardial regeneration. In addition, we will briefly review recent clinical trials, controversies and unresolved issues.

Stem Cells can be Used to Replace Lost and Damaged Cells

Replacement of dead or dysfunctional cardiac myocytes through cell-based therapies represents a novel option for the treatment of cardiovascular diseases. Stem cell-based therapies may achieve true cardiac regeneration by renewing the pool of cardiomyocytes or may lead to cardiac repair by stimulating neovascularization, cytoprotection, modulation of inflammation and by extracellular matrix deposition (Laflamme and Murry 2005, 2011).

There are several approaches for the use of stem cells in cardiovascular therapies. The primary strategy is cell therapy that increases the number of functional cardiomyocytes in the infarcted myocardium by cell transplantation, which entails the delivery of a volume or mass of cells to the damaged tissue. The implanted cells

A. Itzhaki-Alfia (✉)
Tamman Cardiovascular Research Institute, Leviev Heart Center,
Sheba Medical Center, 52621 Tel-Hashomer, Israel
e-mail: itzhaki.ayelet@gmail.com

J. Leor
Tamman Cardiovascular Research Institute,
Leviev Heart Center, Tel-Aviv University, Tel-Hashomer, Israel

Neufeld Cardiac Research Institute, Sackler Faculty of Medicine,
Tel-Aviv University, Tel-Aviv, Israel

J. Hescheler, E. Hofer (eds.), *Adult and Pluripotent Stem Cells*,
DOI 10.1007/978-94-017-8657-7_2,

may have the potential for growth, self-repair and remodeling by forming new cardiac tissue (Etzion et al. 2001; Leor et al. 1996). Such an approach can be divided into two treatment strategies: the first uses differentiated cells to replace the scar tissue with living cells, while the second uses undifferentiated stem cells as building blocks to grow specific types of tissue in situ, in an attempt to regenerate the myocardium (Penn et al. 2002; Zimmet and Krum 2008). Stem cells could be used to replace lost or damaged cells, if they possess the capacity to produce a large number of specific cell types in situ or ex vivo for transplantation.

Adult Stem Cells are Involved in Tissue Replacement, Repair and Maintenance

Adult stem cells, which can be found in adult tissue, are involved in tissue replacement, repair and maintenance, and are already committed to a specific developmental pathway. Usually they generate cell types already present in the same tissue from which they are derived. These cells are multi-potent and have the potential to develop into several, but not all cells types (Passier and Mummery 2003). Adult stem cells include bone marrow cells (Orlic et al. 2001), haemopoietic stem cells (HSC) (Muller-Sieburg et al. 2002), endothelial progenitor cells (EPCs) (Asahara et al. 1999), mesenchymal stem cells (MSCs), and resident cardiac stem cells (Smith et al. 2007; Messina et al. 2004; Oh et al. 2003; Urbanek et al. 2003).

Cells for Cardiovascular Regenerative Therapy

The optimal cells for human myocardial regeneration should be autologous, capable of differentiating into all adult heart cell types, without forming tumors. Various cells proposed to be beneficial for myocardial regeneration are shown in Table 2.1.

Compared with adult stem cells, human embryonic (hES) stem cells and induced pluripotent stem cells (iPS) are more flexible and versatile, and can differentiate into almost every type of cell in the human body. They are capable of almost unlimited division and proliferation, can be grown fairly easily and produce large numbers of stem cells *in vitro*. In contrast, most adult stem cells do not multiply readily after isolation, which is a major problem in stem cell replacement therapies. Since adult stem cells can differentiate into only a limited number of cell types, their versatility for therapy is limited. Within this property, however, lies a major advantage: such cells are committed and do not form tumors. Adult stem cells have another advantage over ES cells in cardiovascular regenerative medicine (see Table 2.1). They are autologous and do not carry the risk of rejection or the need for immunosuppressive therapy connected with adverse effects. However, all of these cell sources and their

Table 2.1 Advantages and limitations of various cell sources for myocardial repair

	Autologous	Easily obtainable	Highly expandable	Cardiac myogenesis	Clinical experience	Safety concerns
Human embryonic stem cells	No	No	Yes	Yes	No	Yes
Mesenchymal stem cells	Yes	No	Depends on age	?	Yes	No
Crude bone marrow cells	Yes	Yes	Depends on age	?	Yes	No
Cardiac stem cells	Yes	No	Depends on the method	Yes	Yes	?
iPS	Yes	No	No	Yes	No	Yes

mechanisms need further research to achieve improvements in therapies. Despite significant advances in stem cell biology, the best cell type and delivery strategy remains to be defined. It is possible that different diseases require different cells for optimal therapy.

Human Heart Regeneration

The question of whether human hearts are able to regenerate has been the subject of furious debate for several years. However, data gathered in the last decade have suggested that the human heart has self-renewal potential. For example, in hearts of patients who died after myocardial infarction (MI), 4 and 1 % of myocytes, in the border zone and remote area, are positive for the cell proliferation marker Ki-67 (Beltrami et al. 2001). Some Ki-67-positive cells, however, were also present in control hearts, suggesting that myocyte proliferation not only occurs during life but is further enhanced following injury (Beltrami et al. 2001). It has been estimated that muscle cells of the entire heart are replaced every 4.5 years (Anversa et al. 2006).

Another method to estimate the turnover of human cardiomyocytes utilized ^{14}C levels in the nucleus as a age biomarker (Bergmann et al. 2009). Due to nuclear testing, atmospheric levels of ^{14}C have been increased, and ^{14}C has spread around the world as $^{14}CO_2$ However, following the nuclear test ban in 1963, atmospheric ^{14}C levels gradually declined. Because ^{14}C is incorporated into the DNA of dividing cells, and DNA remains stable after its last division, ^{14}C concentration in DNA mirrors the atmospheric levels of ^{14}C at the time of its last division, and therefore can be used to determine the age of human cells (Spalding et al. 2005). Mathematical calculations estimated that ~1 and ~0.4 % of myocytes are renewed every year at the ages of 20 and 80, leading to the estimation that ~45 % of myocytes are replaced during a normal human life span (Spalding et al. 2005; Bergmann et al. 2009)

The reason for the difference between the two studies is unclear. Ki-67 labeling may over-estimate cardiomyocyte proliferation because cardiomyocytes sometimes undergo DNA synthesis without nuclear division, or nuclear division without cytokinesis. It is also unclear from these studies whether the newly formed myocytes are derived from pre-existing myocytes or from human cardiac progenitor cells (hCPCs). Genetic fate-mapping studies in mice have suggested that cardiomyocyte renewal following myocardial injury originate from either resident CPCs, (Hsieh et al. 2007) or cardiomyocyte dedifferentiation and proliferation (Porrello et al. 2011; Kubin et al. 2011). Nonetheless, recent human specimen studies provide evidence that the human heart retains self-renewal capacity, but with a low frequency.

Human Resident Cardiac Stem Cells for Myocardial Regeneration and Repair

The discovery that the adult mammalian heart contains a small pool of stem cells has generated a new field in cardiac regenerative medicine. These specific subpopulations of hCPCs are involved in cell replacement, repair and maintenance of the myocardium, cardiac cellular homeostasis during aging and possess stem cell characteristics.

Surprisingly, different researchers have described different markers and phenotypes of hCPCs (Table 2.2). These markers and cell types include c-Kit-positive cells (CD117, the receptor for stem cell factor) (Bearzi et al. 2007), cardiospheres (Messina et al. 2004), cardiosphere-derived cells (Smith et al. 2007), stem cell antigen (Sca)-1-positive cells (Morrison et al. 1997), islet 1-positive cells (Laugwitz et al. 2005), stage-specific embryonic antigen 1-positive cells (Pouly et al. 2008; Ott et al. 2007) and Wt1 epicardial-derived cells (Zhou et al. 2008).

A. P. Beltrami (Beltrami et al. 2003, 2001) from the Anversa group was the first to isolate and grow CPCs from rat heart. C. Bearzi (Bearzi et al. 2007) from the Anversa group was the first to report that human cardiac stem cells (hCSCs), isolated from human biopsies, were positive for c-kit, a stem cell growth factor receptor, also known as CD117. The c-kit receptor was originally detected in murine HSCs with ability to restore bone marrow in irradiated recipients (Orlic et al. 1993). Anversa et al. showed a major contribution of c-kit progenitor cells to cardiomyogenesis during embryonic development (Ferreira-Martins et al. 2012), aging (Tang et al. 2010), and after myocardial infarction (Beltrami et al. 2003). Furthermore, c-kit is re-expressed in dedifferentiated cardiomyocytes during myocardial regeneration (Kubin et al. 2011)

Another proposed marker of CPC population is the insulin gene enhancer protein Islet-1, a marker of the secondary heart field that is involved in embryonic heart formation, gives rise to the outflow tract and ventricles. K.-L. Laugwitz (Laugwitz et al. 2005) was the first to find cardioblasts, Islet-1+ progenitor cells, isolated from

Table 2.2 Human resident cardiac stem cells markers and cell isolation methods

Group	Positive markers	Negative marker	Isolation methods	Source of tissues
(Messina et al. 2004)	c-kit$^+$, CD34$^+$, SCA-1$^+$, KDR$^+$		Trypsin and 0.1 % Collagenase IV	Atrial or ventricular biopsy specimens
(Laugwitz et al. 2005)	islet-1$^+$, Nkx2.5$^+$, GATA4$^+$	sca1$^-$, CD31$^-$, c-kit$^-$	Collagenase II	Right atrial tissue pediatric surgery
(Bearzi et al. 2007)	c-kit$^+$	CD45$^-$, CD34$^-$, CD31$^-$, KDR$^-$	Collagenase IV	–
(Smith et al. 2007)	c-kit$^+$, CD105$^+$, CD90$^+$, CD34$^+$, CD31$^+$	MDR1$^-$, CD133$^-$, CD45$^-$	Tissue fragments were cultured as "explants"	Percutaneous right ventricular biopsy
(Goumans et al. 2008)	CD105$^+$, SCA-1$^+$	CD45$^-$, CD34$^-$, CD133$^-$, CD14$^-$	Collagenase A	Fetal tissue and atrial biopsies
(Itzhaki-Alfia et al. 2009)	c-kit$^+$, GATA4$^+$, Ki-67$^+$	CD45$^-$, CD31$^-$, Collagen$^-$, CD68$^-$, Tryptase$^-$	Trypsin-EDTA 0.25 % and dispase II	Myocardial tissue from right or left atrial appendages, right ventricle or left ventricle

pediatric surgery. Those cells were cultured on cardiac mesenchymal feeder layers and were positive to early cardiac mesoderm markers (Nkx2.5 and GATA-4). Finally, while previous groups described cultured CPCs as adherent cells, Messina et al. described human cardiospheres: suspension of cells obtained from human myocardial biopsy samples (Messina et al. 2004). Thus, the markers that phenotype the hCPC population are varied among different researchers.

The anatomical source of hCPCs in the heart is an important factor in understanding their role in maintaining cardiac homeostasis and therapeutic potential. Pouly et al. conducted a histopathological study analyzing tissue samples from biopsies of a right-sided septum from heart-transplanted patients, and right atrial appendage specimens from patients with ischemic cardiomyopathy (Pouly et al. 2008). They found higher amounts of c-kit$^+$ cells in septal biopsies compared with those of the right atrial appendage. However, the septal biopsies were taken from heart transplant patients, a third of whom had some degree of rejection and were on immunosuppressive therapy. When we compared cell cultures from tissue derived from various patients and analyzed the data, multi-variant models were incorporated to eliminate potential differences in baseline characteristics. This showed that the number of cultured c-kit$^+$ cells produced from the right atrium was 5 fold higher than from other sources (Itzhaki-Alfia et al. 2009).

An Efficient and Reproducible Enzymatic Protocol to Isolate HCPCs

Each group in the field worked so far with a different isolation method and different tissue sources. The cell yield and cluster formation were different with each of the techniques used for cell isolation. Methods to isolate hCPCs include enzymatic (Goumans et al. 2008; Laugwitz et al. 2005; Bearzi et al. 2007; Argentin et al. 1994; Parker et al. 1990a, b; Zhou et al. 2000; Itzhaki-Alfia et al. 2009) and non-enzymatic approaches (Smith et al. 2007; Messina et al. 2004). Using the enzymatic approach, researchers used various enzymes to dissociate the cells from the myocardial specimens. Then, stem cells were isolated from the dissociated cells with a magnetic cell sorter using antibodies against stem cell markers, such as c-kit (Bearzi et al. 2007). However, the cell yield was low and required several months to grow a sufficient number of cells for potential therapeutic applications (Argentin et al. 1994; Parker et al. 1990a, b; Zhou et al. 2000; Forrester et al. 2009). In the non-enzymatic method, tissue samples were seeded, allowing for the spontaneous outgrowth of cells which could then be harvested as cell aggregates in suspension, e.g. cardiospheres (Messina et al. 2004).

We improved the technology and developed an enzymatic digestion cocktail as well as an efficient and reproducible protocol to isolate c-kit$^+$ hCPCs from diverse tissue samples (Itzhaki-Alfia et al. 2009). Compared with other methods (Argentin et al. 1994; Parker et al. 1990a, b; Zhou et al. 2000; Table 2.3), our method has a near 100% success rate in isolating hCPCs (Itzhaki-Alfia et al. 2009). During

Table 2.3 Isolation methods used to dissociate human heart tissue

Technique used	Researchers	Percentage of tissue that yielded progenitor cells	Living/dead cell ratios	The amount of viable cells per gram tissue
RDB proteolytic enzyme (Institute of Biology, Nes-Ziona,Israel)	(Shneyvays et al. 1998)	No data	No data	No data
Collagenase	(Bearzi et al. 2007)	8 of 12	No data	No data
Explanted technique	(Bearzi et al. 2007)	46 of 70	No data	6,000
Explanted technique	(Smith et al. 2007)	69 of 70	No data	No data
0.1 % trypsin, 0.1 % collagenase, 0.025 % DNAase	(Parker et al. 1990b)	No data	No data	No data
Trypsin-EDTA 0.25 % and dispase II	(Itzhaki-Alfia et al. 2009)	249 of 249 samples	95–98 %	$7 \times 10^6 \pm 6.53 \times 10^5$/g

our study we compared our method with different other methods, enzymatic digestion cocktails and isolation protocols with respect to cell yield. Isolation methods described in Table 2.3 were tested on human tissues, whereas these methods were originally used to obtain cells from rodents, a fact which might explain the low yield compared with our method.

Patient Characteristics and the Number of hCPCs

One of the most important challenges in the field of regenerative medicine is to predict the yield of hCSC production from a specific patient. Previous reports correlated increased numbers of CPCs in humans with aortic stenosis, acute and chronic ischemia, and cardiomyopathies (Beltrami et al. 2001, 2003; Kajstura et al. 1998; Urbanek et al. 2005b; Kubo et al. 2008). Together, these findings, which are consistent with ours, support the notion that an endogenous reparative system is stimulated in the injured heart (Hsieh et al. 2007).

Several studies have attempted to determine the correlation between patient characteristics and hCPC yield. While it has been found that coronary artery disease does not affect CPC yield (Aghila Rani et al. 2009) the use of ß-blockers was positively associated with CPC percentage at first passage (Gambini et al. 2012).

We found that patients with hypertension had 2-fold greater percentage of c-kit$^+$ cells in culture, compared with normotensive patients. No correlation was found between medical therapy, including aspirin, ACE inhibitors, amiodarone or statins and the number of c-kit$^+$ cells (Itzhaki-Alfia et al. 2009). Another intriguing finding was that female gender is associated with a higher number of c-kit$^+$ cells (Itzhaki-Alfia et al. 2009). Although the explanation for this finding is unclear, it is consistent with previous reports suggesting variability in stem cell function and numbers between genders (Olivetti et al. 1995; Anversa et al. 2005; Nelson et al. 2007).

Previous work on stem cell aging in the adult heart has indicated that the cardiac progenitor population undergoes an aging process, which could be attributed to either telomere shortening-mediated cell senescence or cumulative cellular trauma (Rota et al. 2008; Anversa et al. 2005; Sussman and Anversa 2004). Consequently, histological analysis of myocardial tissue from elderly patients with cardiomyopathy showed lower numbers of c-kit^{+} cells compared with younger patients (Chimenti et al. 2003). It has been suggested that in older patients, stem cell isolation from cardiac biopsies may not succeed, and such cells may not be available for cell therapy (Aghila Rani et al. 2009). We found however, that age does not significantly affect the number of hCPCs derived from myocardial specimens (Itzhaki-Alfia et al. 2009), a finding consistent with other researchers who suggest that chronological age is not necessarily the main determinant of hCPC functionality and frequency (Rota et al. 2008). There are probably other background conditions that play an important role in the physiology of hCPCs. An additional explanation is that in the aged heart, hCPCs are probably located in certain regions within the myocardium (e.g. the atrium) that retain significant amounts of telomerase-competent hCPCs with long telomeres, which replicate in response to pathological states and participate in regeneration of damaged areas (Gonzalez et al. 2008).

Cardiac Progenitor Cells for Myocardial Repair

Stimulation of endogenous regenerative activity is an attractive strategy for myocardial regeneration and repair. Regeneration can be achieved by stimulation of self-renewal of cardiomyocytes or hCPCs in situ, with a drug or cytokine (Bock-Marquette et al. 2004; Smart et al. 2007; Ruvinov et al. 2011). Alternatively, regeneration can be achieved by *ex vivo* propagation of hCPCs followed by transplantation of the cells into the injured area (Fig. 2.1; Bearzi et al. 2007; Beltrami et al. 2003; Quaini et al. 2004).

Regeneration can be achieved by *ex vivo* proliferation of hCPCs. Small biopsies of the myocardium can be digested by a proteolysis enzyme cocktail, followed by proliferation and expansion in culture. In the final stage, these cells are transplanted into the injured area. Regeneration can be also achieved by stimulation of endogenous cardiomyocytes or hCPCs *in situ*, by delivering to injured myocardium specific growth factors or cytokines.

Animal studies have suggested that transplantation of cells expressing c-kit into the heart could stimulate formation of new blood vessels and myocardium (Beltrami et al. 2003; Tang et al. 2010). Suspension culture of cells obtained from human myocardial biopsy samples yields cardiospheres that contain a mixed population

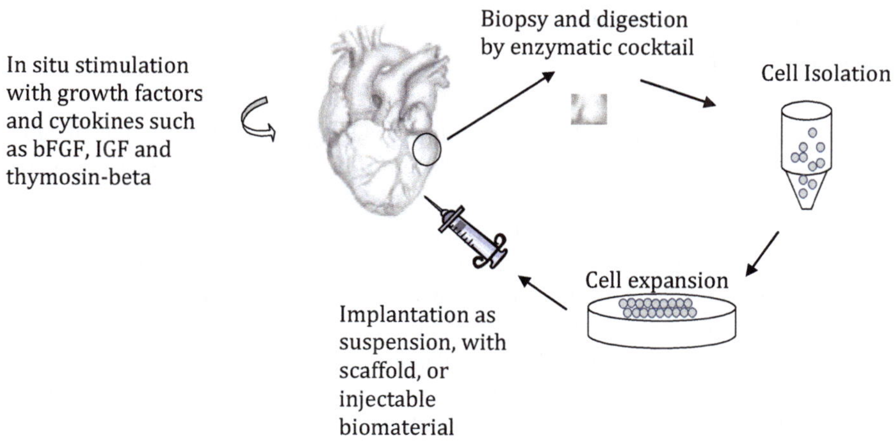

Fig. 2.1 Cardiac progenitor cells for myocardial regeneration and repair

of c-kit⁻ and Sca-1-expressing cells that could stimulate cardiac regeneration (Chimenti et al. 2010).

Whether CSCs are superior to MSCs for myocardial regeneration is unknown. However, a recent study found that hCSCs have a greater ability to engraft and improve cardiac function compared with MSCs (Oskouei et al. 2012). When both hCSCs and MSCs were separately injected into a mouse MI model, both cell types improved ejection fraction, but MSCs required a higher dose (1×10^6) compared with CSCs (3.6×10^4 cells). Moreover, CSCs were superior in reducing scar tissue. These results suggest enhanced engraftment of CSCs compared with MSCs (Oskouei et al. 2012).

To become a viable clinical option, cardiac cell therapy should adopt improved methods of cell engraftment and survival. These improvements can be done by the addition of growth factors (Takehara et al. 2008; Penn and Mangi 2008), genetic engineering (Mohsin et al. 2012), and restoration of the damaged extracellular matrix with scaffolds or injectable biomaterials (Ruvinov et al. 2011; Landa et al. 2008). Alternatively, resident CPCs could be activated in situ by delivering specific growth factors or cytokines to injured myocardium. For example, local delivery of biotinylated IGF-1 complex increased cardiomyocyte growth both *in vitro* and *in vivo* (Davis et al. 2006; D'Amario et al. 2011). Another approach for CSC stimulation was achieved by co-culturing them together with MSCs (Hatzistergos et al. 2010). The number of c-kit CSCs was then 6-fold higher than in CSCs cultured without MSCs (Hatzistergos et al. 2010).

From Bench to Bedside: hCSCs and Cardiospheres in Clinical Trials

The encouraging results of CPCs in animal models led to the initiation of "first in man" clinical trials: ALCADIA (NCT00981006), SCIPIO (NCT00474461), and CADUCEUS (NCT00893360). Preliminary results have recently been published from the SCIPIO and CADUCEUS trials (Makkar et al. 2012; Bolli et al. 2011). Both were designed to assess the feasibility and safety of intracoronary injection of autologous hCPCs (1 million in the SCIPIO, and 12.5–25 million cells in the CADUCEUS) after recent infarction. The SCIPIO study (Bolli et al. 2011) involved c-kit-expressing cells cultured from explanted atrial tissue, whereas the CADUCEUS study (Makkar et al. 2012) involved cardiospheres cultured from biopsies obtained from inter-ventricular septum. Both studies reported on safety and reduction in myocardial scar mass following cell treatment, but only the SCIPIO trial reported improvement in left ventricular ejection fraction. These preliminary results should be viewed with caution, since the number of patients in the treatment group of each study was small (16 in SCIPIO and 17 in CADUCEUS), and neither study included a true placebo group. Many questions are raised following these preliminary trials. Nevertheless, these trials on hCPC treatment for ischemic heart failure established a milestone in the history of cardiac cell therapy.

Unresolved Issues

Substantial gaps remain in our knowledge regarding hCPCs and the mechanisms by which they might promote regeneration. First, the existence and regenerative potential of hCPCs in the adult heart has been questioned and debated. Second, no consensus exists relating to phenotype definition or isolation technique of these populations, many of which have not been compared. Third, genetic fate mapping, a stringent scientific technique used to establish the derivatives of a progenitor cell population, suggests that c-kit-positive cells and cardiospheres do not differentiate into cardiomyocytes (Zaruba et al. 2010). In mice, c-kit-positive cells derived from bone marrow stimulate cardiomyocyte renewal after infarction without direct transdifferentiation (Loffredo et al. 2011). This effect might be attributable to paracrine factors released by the implanted cells on a distinct population of endogenous cardiac cells (Hatzistergos et al. 2010). While advanced age and heart disease did not affect the number of hCPCs (Itzhaki-Alfia et al. 2009), the effect on hCPC function and reparative properties is less clear. Finally, the adoptive transfer of hCSCs results in modest repair due in part to lack of survival, proliferation, and commitment of the transplanted cells after MI.

Summary and Perspective

The area of resident hCPCs is still in its infancy and significant amounts of data regarding the characterization and function of these cells are lacking. However, data from several labs show that the adult human heart, especially the right atrium, retains a unique cell population with stem cell properties. These cells can be isolated, expanded and stored, and could be a viable option to treat patients with heart disease. Augmentation of endogenous regenerative activity in injured hearts will open exciting new options in cardiovascular regenerative medicine, and revolutionize our approach to patients with end-stage heart failure. Larger clinical trials, powered to show clinically meaningful outcomes, are warranted to prove the safety and efficacy of these strategies.

Source of Funding This work was supported by an EC grant (FP7 HEALTH-F2–2009-222995 INELPY) to JL.

References

Aghila Rani KG, Jayakumar K, Sarma PS, Kartha CC (2009) Clinical determinants of ckit-positive cardiac cell yield in coronary disease. Asian Cardiovasc Thorac Ann 17(2):139–142. doi:10.1177/0218492309103292

Anversa P, Rota M, Urbanek K, Hosoda T, Sonnenblick E, Leri A, Kajstura J, Bolli R (2005) Myocardial aging. Basic Res Cardio 100(6):482–493

Anversa P, Kajstura J, Leri A, Bolli R (2006) Life and death of cardiac stem cells: a paradigm shift in cardiac biology. Circulation 113(11):1451–1463. doi:113/11/1451 [pii] 10.1161/CIRCULATIONAHA.105.595181

Argentin S, Ardati A, Tremblay S, Lihrmann I, Robitaille L, Drouin J, Nemer M (1994) Developmental stage-specific regulation of atrial natriuretic factor gene transcription in cardiac cells. Mol Cell Biol 14(1):777–790

Asahara T, Masuda H, Takahashi T, Kalka C, Pastore C, Silver M, Kearne M, Magner M, Isner JM (1999) Bone marrow origin of endothelial progenitor cells responsible for postnatal vasculogenesis in physiological and pathological neovascularization. Circ Res 85(3):221–228. doi:10.1161/01.res.85.3.221

Bearzi C, Rota M, Hosoda T, Tillmanns J, Nascimbene A, De Angelis A, Yasuzawa-Amano S, Trofimova I, Siggins RW, LeCapitaine N, Cascapera S, Beltrami AP, D'Alessandro DA, Zias E, Quaini F, Urbanek K, Michler RE, Bolli R, Kajstura J, Leri A, Anversa P (2007) Human cardiac stem cells. Proc Natl Acad Sci U S A 104(35):14068–14073. doi:10.1073/pnas.0706760104

Beltrami AP, Urbanek K, Kajstura J, Yan S-M, Finato N, Bussani R, Nadal-Ginard B, Silvestri F, Leri A, Beltrami CA, Anversa P (2001) Evidence that human cardiac Myocytes divide after myocardial infarction. N Engl J Med 344(23):1750–1757. doi:10.1056/nejm200106073442303

Beltrami AP, Barlucchi L, Torella D, Baker M, Limana F, Chimenti S, Kasahara H, Rota M, Musso E, Urbanek K, Leri A, Kajstura J, Nadal-Ginard B, Anversa P (2003) Adult cardiac stem cells are multipotent and support myocardial regeneration. Cell 114(6):763–776. doi:S0092867403006871 [pii]

Bergmann O, Bhardwaj RD, Bernard S, Zdunek S, Barnabe-Heider F, Walsh S, Zupicich J, Alkass K, Buchholz BA, Druid H, Jovinge S, Frisen J (2009) Evidence for cardiomyocyte renewal in humans. Science 324(5923):98–102. doi:10.1126/science.1164680

Bock-Marquette I, Saxena A, White MD, Dimaio JM, Srivastava D (2004) Thymosin beta4 activates integrin-linked kinase and promotes cardiac cell migration, survival and cardiac repair. Nature 432(7016):466–472. doi:10.1038/nature03000

Bolli R, Chugh AR, D'Amario D, Loughran JH, Stoddard MF, Ikram S, Beache GM, Wagner SG, Leri A, Hosoda T, Sanada F, Elmore JB, Goichberg P, Cappetta D, Solankhi NK, Fahsah I, Rokosh DG, Slaughter MS, Kajstura J, Anversa P (2011) Cardiac stem cells in patients with ischaemic cardiomyopathy (SCIPIO): initial results of a randomised phase 1 trial. Lancet 378(9806):1847–1857. doi:10.1016/S0140-6736(11)61590-0

Chien KR (2005) Alchemy and the new age of cardiac muscle cell biology. PLoS Biol 3(4):e131

Chimenti C, Kajstura J, Torella D, Urbanek K, Heleniak H, Colussi C, Di Meglio F, Nadal-Ginard B, Frustaci A, Leri A, Maseri A, Anversa P (2003) Senescence and death of primitive cells and myocytes lead to premature cardiac aging and heart failure. Circ Res 93(7):604–613. doi:10.1161/01.res.0000093985.76901.af

Chimenti I, Smith RR, Li TS, Gerstenblith G, Messina E, Giacomello A, Marban E (2010) Relative roles of direct regeneration versus paracrine effects of human cardiosphere-derived cells transplanted into infarcted mice. Circ Res 106(5):971–980. doi:10.1161/CIRCRESAHA.109.210682

D'Amario D, Cabral-Da-Silva MC, Zheng H, Fiorini C, Goichberg P, Steadman E, Ferreira-Martins J, Sanada F, Piccoli M, Cappetta D, D'Alessandro DA, Michler RE, Hosoda T, Anastasia L, Rota M, Leri A, Anversa P, Kajstura J (2011) Insulin-like growth factor-1 receptor identifies a pool of human cardiac stem cells with superior therapeutic potential for myocardial regeneration. Circ Res 108(12):1467–1481. doi:10.1161/CIRCRESAHA.111.240648

Davis ME, Hsieh PC, Takahashi T, Song Q, Zhang S, Kamm RD, Grodzinsky AJ, Anversa P, Lee RT (2006) Local myocardial insulin-like growth factor 1 (IGF-1) delivery with biotinylated peptide nanofibers improves cell therapy for myocardial infarction. Proc Natl Acad Sci U S A 103(21):8155–8160. doi:0602877103 [pii] 10.1073/pnas.0602877103

Etzion S, Battler A, Barbash IM, Cagnano E, Zarin P, Granot Y, Kedes LH, Kloner RA, Leor J (2001) Influence of embryonic cardiomyocyte transplantation on the progression of heart failure in a rat model of extensive myocardial infarction. J Mol Cell Cardiol 33(7):1321–1330

Ferreira-Martins J, Ogorek B, Cappetta D, Matsuda A, Signore S, D'Amario D, Kostyla J, Steadman E, Ide-Iwata N, Sanada F, Iaffaldano G, Ottolenghi S, Hosoda T, Leri A, Kajstura J, Anversa P, Rota M (2012) Cardiomyogenesis in the developing heart is regulated by c-kit-positive cardiac stem cells. Circ Res 110(5):701–715. doi:10.1161/CIRCRESAHA.111.259507

Forrester JS, Makkar RR, Marban E (2009) Long-term outcome of stem cell therapy for acute myocardial infarction: right results, wrong reasons. J Am Coll Cardiol 53(24):2270–2272. doi:S0735-1097(09)01063-8 [pii] 10.1016/j.jacc.2009.03.023

Gambini E, Pesce M, Persico L, Bassetti B, Gambini A, Alamanni F, Agrifoglio M, Capogrossi MC, Pompilio G (2012) Patient profile modulates cardiac c-kit$^+$ progenitor cell availability and amplification potential. Transl Res 160(5):1–11

Gonzalez A, Rota M, Nurzynska D, Misao Y, Tillmanns J, Ojaimi C, Padin-Iruegas ME, Muller P, Esposito G, Bearzi C, Vitale S, Dawn B, Sanganalmath SK, Baker M, Hintze TH, Bolli R, Urbanek K, Hosoda T, Anversa P, Kajstura J, Leri A (2008) Activation of cardiac progenitor cells reverses the failing heart senescent phenotype and prolongs lifespan. Circ Res 102(5):597–606. doi:10.1161/circresaha.107.165464

Goumans M-J, de Boer TP, Smits AM, van Laake LW, van Vliet P, Metz CHG, Korfage TH, Kats KP, Hochstenbach R, Pasterkamp G, Verhaar MC, van der Heyden MAG, de Kleijn D, Mummery CL, van Veen TAB, Sluijter JPG, Doevendans PA (2008) TGF-[beta]1 induces efficient differentiation of human cardiomyocyte progenitor cells into functional cardiomyocytes in vitro. Stem Cell Res 1(2):138–149

Hatzistergos KE, Quevedo H, Oskouei BN, Hu Q, Feigenbaum GS, Margitich IS, Mazhari R, Boyle AJ, Zambrano JP, Rodriguez JE, Dulce R, Pattany PM, Valdes D, Revilla C, Heldman

AW, McNiece I, Hare JM (2010) Bone marrow mesenchymal stem cells stimulate cardiac stem cell proliferation and differentiation. Circ Res 107(7):913–922. doi:10.1161/CIRCRESAHA.110.222703

Hsieh PCH, Segers VFM, Davis ME, MacGillivray C, Gannon J, Molkentin JD, Robbins J, Lee RT (2007) Evidence from a genetic fate-mapping study that stem cells refresh adult mammalian cardiomyocytes after injury. Nat Med 13(8):970–974

Itzhaki-Alfia A, Leor J, Raanani E, Sternik L, Spiegelstein D, Netser S, Holbova R, Pevsner-Fischer M, Lavee J, Barbash IM (2009) Patient characteristics and cell source determine the number of isolated human cardiac progenitor cells. Circulation 120(25):2559–2566. doi:CIRCULATIONAHA.109.849588 [pii] 10.1161/CIRCULATIONAHA.109.849588

Kajstura J, Leri A, Finato N, Di Loreto C, Beltrami CA, Anversa P (1998) Myocyte proliferation in end-stage cardiac failure in humans. Proc Natl Acad Sci U S A 95(15):8801–8805. doi:10.1073/pnas.95.15.8801

Kubin T, Poling J, Kostin S, Gajawada P, Hein S, Rees W, Wietelmann A, Tanaka M, Lorchner H, Schimanski S, Szibor M, Warnecke H, Braun T (2011) Oncostatin M is a major mediator of cardiomyocyte dedifferentiation and remodeling. Cell Stem Cell 9(5):420–432. doi:S1934-5909(11)00394-8 [pii] 10.1016/j.stem.2011.08.013

Kubo H, Jaleel N, Kumarapeli A, Berretta RM, Bratinov G, Shan X, Wang H, Houser SR, Margulies KB (2008) Increased cardiac myocyte progenitors in failing human hearts. Circulation 118(6):649–657. doi:10.1161/circulationaha.107.761031

Laflamme MA, Murry CE (2005) Regenerating the heart. Nat Biotech 23(7):845–856

Laflamme MA, Murry CE (2011) Heart regeneration. Nature 473(7347):326–335. doi:10.1038/nature10147

Landa N, Miller L, Feinberg MS, Holbova R, Shachar M, Freeman I, Cohen S, Leor J (2008) Effect of injectable alginate implant on cardiac remodeling and function after recent and old infarcts in rat. Circulation 117(11):1388–1396. doi:10.1161/circulationaha.107.727420

Laugwitz K-L, Moretti A, Lam J, Gruber P, Chen Y, Woodard S, Lin L-Z, Cai C-L, Lu MM, Reth M, Platoshyn O, Yuan JXJ, Evans S, Chien KR (2005) Postnatal isl1+ cardioblasts enter fully differentiated cardiomyocyte lineages. Nature 433(7026):647–653

Leor J, Patterson M, Quinones MJ, Kedes LH, Kloner RA (1996) Transplantation of fetal myocardial tissue into the infarcted myocardium of rat. A potential method for repair of infarcted myocardium? Circulation 94(9 Suppl):II332–336

Loffredo FS, Steinhauser ML, Gannon J, Lee RT (2011) Bone marrow-derived cell therapy stimulates endogenous cardiomyocyte progenitors and promotes cardiac repair. Cell Stem Cell 8(4):389–398. doi:10.1016/j.stem.2011.02.002

Makkar RR, Smith RR, Cheng K, Malliaras K, Thomson LE, Berman D, Czer LS, Marban L, Mendizabal A, Johnston PV, Russell SD, Schuleri KH, Lardo AC, Gerstenblith G, Marban E (2012) Intracoronary cardiosphere-derived cells for heart regeneration after myocardial infarction (CADUCEUS): a prospective, randomised phase 1 trial. Lancet 379(9819):895–904. doi:10.1016/S0140-6736(12)60195-0

Messina E, De Angelis L, Frati G, Morrone S, Chimenti S, Fiordaliso F, Salio M, Battaglia M, Latronico MVG, Coletta M, Vivarelli E, Frati L, Cossu G, Giacomello A (2004) Isolation and expansion of adult cardiac stem cells from human and murine heart. Circ Res 95(9):911–921. doi:10.1161/01.res.0000147315.71699.51

Mohsin S, Khan M, Toko H, Bailey B, Cottage CT, Wallach K, Nag D, Lee A, Siddiqi S, Lan F, Fischer KM, Gude N, Quijada P, Avitabile D, Truffa S, Collins B, Dembitsky W, Wu JC, Sussman MA (2012) Human cardiac progenitor cells engineered with pim-i kinase enhance myocardial repair. J Am Coll Cardiol 60(14):1278–1287. doi:10.1016/j.jacc.2012.04.047

Morrison SJ, Wandycz AM, Hemmati HD, Wright DE, Weissman IL (1997) Identification of a lineage of multipotent hematopoietic progenitors. Development 124(10):1929–1939

Muller-Sieburg CE, Cho RH, Thoman M, Adkins B, Sieburg HB (2002) Deterministic regulation of hematopoietic stem cell self-renewal and differentiation. Blood 100(4):1302–1309

Nelson WD, Zenovich AG, Ott HC, Stolen C, Caron GJ, Panoskaltsis-Mortari A, Barnes SA III, Xin X, Taylor DA (2007) Sex-dependent attenuation of plaque growth after treatment with bone marrow mononuclear cells. Circ Res 101(12):1319–1327. doi:10.1161/circresaha.107.155564

Oh H, Bradfute SB, Gallardo TD, Nakamura T, Gaussin V, Mishina Y, Pocius J, Michael LH, Behringer RR, Garry DJ, Entman ML, Schneider MD (2003) Cardiac progenitor cells from adult myocardium: Homing, differentiation, and fusion after infarction. Proc Natl Acad Sci U S A 100(21):12313–12318. doi:10.1073/pnas.2132126100

Olivetti G, Giordano G, Corradi D, Melissari M, Lagrasta C, Gambert SR, Anversa P (1995) Gender differences and aging: effects on the human heart. J Am Coll Cardiol 26(4):1068–1079

Orlic D, Fischer R, Nishikawa S, Nienhuis AW, Bodine DM (1993) Purification and characterization of heterogeneous pluripotent hematopoietic stem cell populations expressing high levels of c-kit receptor. Blood 82(3):762–770

Orlic D, Kajstura J, Chimenti S, Jakoniuk I, Anderson SM, Li B, Pickel J, McKay R, Nadal-Ginard B, Bodine DM, Leri A, Anversa P (2001) Bone marrow cells regenerate infarcted myocardium. Nature 410(6829):701–705

Oskouei BN, Lamirault G, Joseph C, Treuer AV, Landa S, Da Silva J, Hatzistergos K, Dauer M, Balkan W, McNiece I, Hare JM (2012) Increased potency of cardiac stem cells compared with bone marrow mesenchymal stem cells in cardiac repair. Stem Cells Transl Med 1(2):116–124. doi:10.5966/sctm.2011-0015

Ott HC, Matthiesen TS, Brechtken J, Grindle S, Goh SK, Nelson W, Taylor DA (2007) The adult human heart as a source for stem cells: repair strategies with embryonic-like progenitor cells. Nat Clin Pract Cardiovasc Med 4(Suppl 1):S27–S39

Parker TG, Chow KL, Schwartz RJ, Schneider MD (1990a) Differential regulation of skeletal alpha-actin transcription in cardiac muscle by two fibroblast growth factors. Proc Natl Acad Sci U S A 87(18):7066–7070

Parker TG, Packer SE, Schneider MD (1990b) Peptide growth factors can provoke "fetal" contractile protein gene expression in rat cardiac myocytes. J Clin Invest 85(2):507–514. doi:10.1172/JCI114466

Passier R, Mummery C (2003) Origin and use of embryonic and adult stem cells in differentiation and tissue repair. Cardiovasc Res 58(2):324–335

Penn MS, Francis GS, Ellis SG, Young JB, McCarthy PM, Topol EJ (2002) Autologous cell transplantation for the treatment of damaged myocardium. Prog Cardiovasc Dis 45(1):21–32

Penn MS, Mangi AA (2008) Genetic enhancement of stem cell engraftment, survival, and efficacy. Circ Res 102(12):1471–1482. doi:102/12/1471 [pii] 10.1161/CIRCRESAHA.108.175174

Porrello ER, Mahmoud AI, Simpson E, Hill JA, Richardson JA, Olson EN, Sadek HA (2011) Transient regenerative potential of the neonatal mouse heart. Science 331(6020):1078–1080. doi:331/6020/1078 [pii] 10.1126/science.1200708

Pouly J, Bruneval P, Mandet C, Proksch S, Peyrard S, Amrein C, Bousseaux V, Guillemain R, Deloche A, Fabiani J-N, Menasch' P (2008) Cardiac stem cells in the real world. J Thorac Cardiovasc Surg 135(3):673–678

Quaini F, Urbanek K, Graiani G, Lagrasta C, Maestri R, Monica M, Boni A, Ferraro F, Delsignore R, Tasca G, Leri A, Kajstura J, Quaini E, Anversa P (2004) The regenerative potential of the human heart. Int J Cardiol 95(Suppl 1):S26–S28. doi:S0167527304900083 [pii]

Rota M, Padin-Iruegas ME, Misao Y, De Angelis A, Maestroni S, Ferreira-Martins J, Fiumana E, Rastaldo R, Arcarese ML, Mitchell TS, Boni A, Bolli R, Urbanek K, Hosoda T, Anversa P, Leri A, Kajstura J (2008) Local activation or implantation of cardiac progenitor cells rescues scarred infarcted myocardium improving cardiac function. Circ Res 103(1):107–116. doi:CIRCRESAHA.108.178525 [pii] 10.1161/CIRCRESAHA.108.178525

Ruvinov E, Leor J, Cohen S (2011) The promotion of myocardial repair by the sequential delivery of IGF-1 and HGF from an injectable alginate biomaterial in a model of acute myocardial infarction. Biomaterials 32(2):565–578. doi:10.1016/j.biomaterials.2010.08.097

Shneyvays V, Nawrath H, Jacobson KA, Shainberg A (1998) Induction of apoptosis in cardiac myocytes by an A3 adenosine receptor agonist. Exp Cell Res 243(2):383–397. doi:S0014-4827(98)94134-9 [pii] 10.1006/excr.1998.4134

Smart N, Risebro CA, Melville AA, Moses K, Schwartz RJ, Chien KR, Riley PR (2007) Thymosin beta-4 is essential for coronary vessel development and promotes neovascularization via adult epicardium. Ann N Y Acad Sci 1112:171–188. doi:annals.1415.000 [pii] 10.1196/annals.1415.000

Smith RR, Barile L, Cho HC, Leppo MK, Hare JM, Messina E, Giacomello A, Abraham MR, Marban E (2007) Regenerative potential of cardiosphere-derived cells expanded from percutaneous endomyocardial biopsy specimens. Circulation 115(7):896–908. doi:10.1161/circulationaha.106.655209

Spalding KL, Bhardwaj RD, Buchholz BA, Druid H, Frisen J (2005) Retrospective birth dating of cells in humans. Cell 122(1):133–143. doi:10.1016/j.cell.2005.04.028

Sussman MA, Anversa P (2004) Myocardial aging and senescence: where have the stem cells gone? Ann Rev Physiol 66(1):29–48. doi:10.1146/annurev.physiol.66.032102.140723

Takehara N, Tsutsumi Y, Tateishi K, Ogata T, Tanaka H, Ueyama T, Takahashi T, Takamatsu T, Fukushima M, Komeda M, Yamagishi M, Yaku H, Tabata Y, Matsubara H, Oh H (2008) Controlled delivery of basic fibroblast growth factor promotes human cardiosphere-derived cell engraftment to enhance cardiac repair for chronic myocardial infarction. J Am Coll Cardiol 52(23):1858–1865

Tang XL, Rokosh G, Sanganalmath SK, Yuan F, Sato H, Mu J, Dai S, Li C, Chen N, Peng Y, Dawn B, Hunt G, Leri A, Kajstura J, Tiwari S, Shirk G, Anversa P, Bolli R (2010) Intracoronary administration of cardiac progenitor cells alleviates left ventricular dysfunction in rats with a 30-day-old infarction. Circulation 121(2):293–305. doi:CIRCULATIONAHA.109.871905 [pii] 10.1161/CIRCULATIONAHA.109.871905

Urbanek K, Quaini F, Tasca G, Torella D, Castaldo C, Nadal-Ginard B, Leri A, Kajstura J, Quaini E, Anversa P (2003) From the cover: intense myocyte formation from cardiac stem cells in human cardiac hypertrophy. PNAS 100(18):10440–10445. doi:10.1073/pnas.1832855100

Urbanek K, Rota M, Cascapera S, Bearzi C, Nascimbene A, De Angelis A, Hosoda T, Chimenti S, Baker M, Limana F, Nurzynska D, Torella D, Rotatori F, Rastaldo R, Musso E, Quaini F, Leri A, Kajstura J, Anversa P (2005a) Cardiac stem cells possess growth factor-receptor systems that after activation regenerate the infarcted myocardium, improving ventricular function and long-term survival. Circ Res 97(7):663–673. doi:10.1161/01.res.0000183733.53101.11

Urbanek K, Torella D, Sheikh F, De Angelis A, Nurzynska D, Silvestri F, Beltrami CA, Bussani R, Beltrami AP, Quaini F, Bolli R, Leri A, Kajstura J, Anversa P (2005b) Myocardial regeneration by activation of multipotent cardiac stem cells in ischemic heart failure. PNAS 102(24):8692–8697. doi:10.1073/pnas.0500169102

Zaruba MM, Soonpaa M, Reuter S, Field LJ (2010) Cardiomyogenic potential of C-kit(+)-expressing cells derived from neonatal and adult mouse hearts. Circulation 121(18):1992–2000. doi:10.1161/CIRCULATIONAHA.109.909093

Zhou YY, Wang SQ, Zhu WZ, Chruscinski A, Kobilka BK, Ziman B, Wang S, Lakatta EG, Cheng H, Xiao RP (2000) Culture and adenoviral infection of adult mouse cardiac myocytes: methods for cellular genetic physiology. Am J Physiol Heart Circ Physiol 279(1):H429–H436

Zhou B, Ma Q, Rajagopal S, Wu SM, Domian I, Rivera-Feliciano J, Jiang D, von Gise A, Ikeda S, Chien KR, Pu WT (2008) Epicardial progenitors contribute to the cardiomyocyte lineage in the developing heart. Nature 454(7200):109–113. doi:nature07060 [pii] 10.1038/nature07060

Zimmet H, Krum H (2008) Using adult stem cells to treat heart failure-fact or fiction? Heart Lung Circ 17(Suppl 4):S48–S54

Chapter 3
Endothelial Progenitor Cells Derived from Cord or Peripheral Blood and Their Potential for Regenerative Therapies

Renate Hofer-Warbinek, Caterina Sturtzel, Karl-Heinz Preisegger and Erhard Hofer

Abstract Cell therapies using stem/progenitor cells hold enormous promise for a wide range of serious and hitherto untreatable diseases. Peripheral blood as well as umbilical cord blood have been demonstrated to be ideal sources of stem/progenitor cells differentiating into endothelial-like cells with high proliferative capacities and the capability to incorporate into hypoxic tissue areas as they occur in ischemic cardiomyopathies. These findings have led to the concept that postnatal vasculogenesis does occur and that neovascularisation in the adult is not solely achieved by sprouting from preexisting mature endothelial cells. In any case, irrespective of the debated origin and normal physiological role of these cells, this opened a new field in tissue regeneration as the cells can be used to trigger neovascularization and the restored supply of oxygen and nutrients is a necessary precondition for the regeneration of any damaged tissue. We will therefore here discuss the occurrence, development and therapeutic potential of postnatal vascular stem and progenitor cells from blood. This will include aspects of our recent work identifying factors characterizing and distinguishing blood-derived endothelial progenitor cells from mature endothelial cells of the vessel wall.

Abbreviations

HSC	Hematopoietic stem cell
EPC	Endothelial progenitor cell
CEC	Circulating endothelial cell
ECFC	Endothelial colony forming cell
BOEC	Blood outgrowth endothelial cell
HUVEC	Human umbilical vein endothelial cells

R. Hofer-Warbinek (✉) · C. Sturtzel · E. Hofer
Department of Vascular Biology and Thrombosis Research, Center of Physiology and Pharmacology, Medical University of Vienna, Schwarzspanierstrasse 17, 1090 Vienna, Austria
e-mail: renate.hofer-warbinek@meduniwien.ac.at

K.-H. Preisegger
Institut für morphologische Analytik und Humangenetik, Graz, Austria

J. Hescheler, E. Hofer (eds.), *Adult and Pluripotent Stem Cells*,
DOI 10.1007/978-94-017-8657-7_3, © Springer Science+Business Media Dordrecht 2014

Concept of Vasculogenesis/Angiogenesis—Neovascularization

Historically vasculogenesis is defined, restricted to the development in the embryo, by de novo formation of a primary vascular plexus from mesodermal progenitor cells. Angiogenesis, however, is characterised as the formation of new capillaries solely by sprouting of resident cells from the vessel wall, as it occurs during growth and after wounding. Neovascularization describes generally the formation of new vessels of any size by any mechanism.

Vasculogenesis in Development: Angioblast—Hemangioblast—Hemogenic Endothelium

The embryonic blood islands in the yolk sac of the embryo are the starting points for the development of the arteriovenous vascular system (Ferkowicz and Yoder 2005). Early data have shown that cells destined to generate the hematopoietic stem cells (HSCs) are situated in the center of the blood islands, whereas cells programmed to become embryonic endothelial progenitor cells (angioblasts) are located in the periphery. Spatial association and the sharing of some antigens led initially to the assumption of a common origin of these cells, termed the "hemangioblast", a bi-potential precursor in the mesoderm giving rise to the hematopoietic and the vascular system (Haar and Ackerman 1971; Park et al. 2005). In addition to proposing the existence of hemangioblasts as a transitional population in the embryo, some studies also provided evidence that hemangioblasts exist in the postnatal stage. Cells positive for CD133, mobilized by GM-CSF into peripheral blood, have been described to give rise to hematopoietic cells as well as endothelial cells when cultured *in vitro* (Gehling et al. 2000). In a leukemia model, both endothelial and hematopoietic cells carried the Bcr/Abl fusion gene supporting their origin from common bone marrow progenitor cells (Gunsilius et al. 2000). Furthermore, several additional studies displayed evidence for the existence of a hemangioblast with expression of the markers CD34 and KDR (Cogle et al. 2004; Pelosi et al. 2002).

However, more recent data from murine embryonic models have clearly shown that blood cell clusters bud off from the aortic wall, in the yolk sac, placenta, and from other major vessels including vitellin and umbilical arteries. In these processes the transcription factor Runx1 seems to play an essential role (Zape and Zovein 2011; Zovein et al. 2008). More advanced technologies, e.g. live imaging (Bertrand et al. 2010; Boisset et al. 2010; Eilken et al. 2009; Kissa and Herbomel 2010), further proved that hematopoietic stem cells (HSCs) emerge from the endothelium in all kind of systems. The data suggest the involvement of the Notch signaling pathway, further including HDAC1 and Runx1, maybe in conjunction with a member of the Hox gene family, HoxA3. Furthermore, crosstalks with the BMP and Wnt signaling pathways are indicated. The exact mechanism how HSCs emanate from the hemogenic endothelium is still unclear. Suggestions range from an endothelial to hematopoietic transition (Kissa and Herbomel 2010) to asymmetric division, a mechanism occurring in stem cell niches (Bertrand et al. 2005; Sasaki et al. 2010).

In conclusion, based on these more recent data, it is very unlikely that definitive hematopoiesis takes place other than from the endothelium which puts the hemangioblast concept into question. However, there might be a link between hemangioblasts and the hemogenic endothelium (Lancrin et al. 2009). The findings of the hemogenic endothelium might therefore be an intriguing source for therapeutic application, however the fact that it is a transient phenomenon, which disappears during maturation prevents immediate translational approaches. To clarify this interrelationship between endothelium and the hematopoietic system remains a challenge for further investigations. Additional informations may be obtained via reprogramming such as in the case of human fibroblasts, where Oct4 expression in combination with cytokine treatment resulted in multilineage blood cell production (Szabo et al. 2010).

Potential Endothelial Progenitors of Hematopoietic Origin from Blood

Postnatal neovascularization results solely from angiogenesis, by sprouting, migration and proliferation of mature endothelial cells from preexisting vessels (Folkman and Shing 1992), which is in contrast to vasculogenesis, a developmental process in the embryo where endothelial progenitor cells, so-called angioblasts, participate in the formation of new blood vessels (Risau et al. 1988). This concept was challenged by Asahara et al. (1997) who described the first isolation of progenitor endothelial cells from peripheral blood. They harvested $CD34^{+}$-enriched mononuclear cells from human peripheral blood and were able to differentiate them into endothelial-like cells on fibrin coated plates. Furthermore, in mouse and rabbit hindlimb ischemia models they observed integration of these cells into or close to CD31 positive capillaries of the injured area. This was independent whether they applied heterologous, homologous or autologous endothelial progenitor cells. Although later studies by other groups questioned the phenotype of these cells and their incorporation into vessels (Hur et al. 2004; Rohde et al. 2006), this work set a hallmark by proposing that cells in the circulating bloodstream can contribute to vessel formation in the adult.

In addition to peripheral blood, endothelial progenitor cells were also described in bone marrow, fetal liver and umbilical cord blood (Murohara et al. 2000). The main hypothesis connected to these reports was that these cells originate from the bone marrow and are released into the peripheral circulation upon specific stimuli.

These progenitor cells were described to possess selective surface markers such as CD34 (Asahara et al. 1997; Popa et al. 2006), CD133 (Gehling et al. 2000) or the combination of CD133 and VEGFR2 (Friedrich et al. 2006; Peichev et al. 2000).

The expression of CD14 and low levels of CD34 was in line with a potential relationship to monocytes/macrophages (Romagnani et al. 2005). Cells derived under specific culture conditions from peripheral blood or bone marrow mononuclear cells (Gulati et al. 2003; Hur et al. 2004; Yoon et al. 2005) were also referred to as endothelial progenitor cells.

The terminology of all these progenitor cells included EPCs (endothelial progenitor cells) and angioblasts, but both terms were not used in a precise way. Therefore, Kovacic (Kovacic et al. 2008) defined in his review more precisely angioblasts as "oligopotent progenitor cells, capable of giving rise to the majority of cell types required for complete new vessel formation", whereas the term EPC describes "a committed (lineage restricted) progenitor cell that gives rise only to endothelial progeny and that is capable of capillogenesis and the repair of preexisting endothelium". For therapeutic purposes to build a new, functional and mature vessel it is predictable that both types are essential.

Another terminology to describe potential endothelial progenitors was blood outgrowth cells, which was first used by Lin et al. (2000). Later on it became evident that dependent on the chronology of their appearance in culture one can distinguish between early and late outgrowth cells. Although their nature is not yet fully understood they show a distinct phenotype (Gehling et al. 2007; Rohde et al. 2006, 2007; Yoder et al. 2007). Early EPCs appear in culture within 7 days (Gulati et al. 2003), exhibit spindle like morphology and resemble a hematopoietic colony forming unit and cannot be expanded for more than 3 weeks. They are positive for CD14, CD45 and CD115 in addition to displaying some VEGFR2 and VE-cadherin, and there is a significant overlap in markers with monocytes/macrophages. To further complicate the situation also T-cells have been identified in so-called early EPCs (Hur et al. 2007; Rohde et al. 2007).

In contrast, late outgrowth EPCs do not appear before 2–3 weeks in culture and have an expansion time of 4–8 weeks (Gulati et al. 2003; Hur et al. 2004; Ishikawa and Asahara 2004; Yoder et al. 2007). Although it was described that they possibly develop from a $CD14^+$ peripheral mononuclear fraction, mature cells do not express CD14 anymore, but expose CD34 and VEGFR2. Both cell types, early and late outgrowth cells bind lectin and show uptake of acetylated low-density lipoprotein cholesterol (Gulati et al. 2003; Hur et al. 2004), characteristics not only for endothelial cells but also monocytes. However, different from early EPCs, late EPCs show tube forming capacity and closely resemble endothelial cells in terms of marker expression (VEGFR2, CD31, VE-cadherin, von Willebrand factor). Initially, both types were thought to derive from $CD14^+$ cells and the data suggested their contribution to vessel formation is of paracrine nature or by action of MMP-12 forming a tunnel in the ischemic tissue and facilitating the growth of a new vessel (Anghelina et al. 2004, 2006; Ishikawa and Asahara 2004; Moldovan et al. 2000).

While the initial studies reported more than 10 years ago suggested that EPCs mobilized from bone marrow would directly contribute to neovascularization in ischemic tissues (Asahara et al. 1997), a number of extensive studies performed in the meantime gave contradictory results. Whereas some reports, in line with the initial reports, described that bone marrow-derived EPCs would contribute to the endothelial layer of neovessels formed in response to hypoxia in areas of ischemia or in a tumor environment (Gao et al. 2008; Jujo et al. 2008; Kerbel et al. 2008; Nolan et al. 2007), other authors using bone marrow transplantation or mobilization in transgenic mice were unable to detect any BM-derived cells in tumor neovessels (Gothert et al. 2004; Purhonen et al. 2008). Based on these reports it appeared that the whole issue is not straightforward and would need further clarification. Since other bone marrow-derived cells like monocytes and macrophages contribute to vessel formation by secreting paracrine factors, some of them express endothelial-like markers and come into close contact with neovessels, maybe in a transient manner, certain of the described reports may have over-interpreted their data and possibly taken monocytic cells for endothelial progenitors.

Endothelial Progenitors from the Vessel Wall: EPC, CEC, ECFC

The use of different marker sets and functional descriptions make it even more complicated to compare the numerous in part contradictory studies. Some of these issues were resolved by work of Yoder (2010). The majority of circulating endothelial cells (CECs) have no proliferative potential, it turned out to be very difficult to enumerate them. However, they are described to serve as markers for certain malignancies (Strijbos et al. 2008). Furthermore, the majority of CECs was ascribed to originate of the vessel wall (Lin et al. 2000). Ingram et al. (2004) developed a specific method to isolate and culture outgrowth endothelial cells in a clonal fashion. They identified a complete hierarchy of these outgrowth cells when they plated single cell colonies from peripheral blood or umbilical cord blood, which they termed endothelial colony forming cells (ECFC). Clones with very high proliferative and self renewal capacity were gained as well as cells with low ratio of proliferation. However, these cells could not be distinguished in respect to their marker profiles.

Especially ECFCs from cord blood are described to form robust *in vivo* human blood vessels at least within a certain time frame, more efficient than ECFCs from peripheral blood (Au et al. 2008; Yoder et al. 2007). The existence of ECFC seems to be species specific, e.g. they could be identified in mammals of large size but are hardly or not at all detectable in the bloodstream of mice. In his review, Yoder (2010) tried to clarify the difference between early outgrowth EPC and ECFC. One possibility to explain the controversial reports, obtained by defining various cell surface antigens and using a variety of colony assays of adherent cells, would be that numerous subsets of cells exist which are involved in the process of tissue repair and neovascularization. One kind of EPC would display the feature to synthesize and deposit an endothelial basement membrane when a new vessel is formed. On the contrary, another kind of progenitors, ECFC, have the ability to form a capillary plexus even in the absence of a preexisting vessel.

Surprisingly, clonogenic cells resembling circulating ECFC were also described in cultures of mature endothelial cells like HUVEC and adult aortic endothelial cells (Ingram et al. 2005a, b) supporting the observation of Schwartz and Benditt, obtained as early as 1976 (Schwartz and Benditt 1976), that in the rat aorta there are clusters of dividing cells within areas of nondividing cells.

Since ECFC cannot be distinguished from endothelial cells that reside in a vessel by analysis of their surface markers, they are difficult to isolate. ECFC express all the markers characteristic for endothelial cells, but in contrast to early EPC in general they expose no CD14, CD45 and CD115 such as hematopoietic stem cells do.

Another source for stem/progenitor cells of endothelial origin bearing mesenchymal characteristics, is based on the ability of endothelial cells of the vessel lining to detach and undergo the socalled "endothelial-to-mesenchymal transition" (Arciniegas et al. 2007; Potenta et al. 2008), which most likely has pathological reasons. This process might involve multipotent, intermediate progenitor cell types and data suggest that the TGFß signaling pathway plays a role.

Stem Cell Niche

The fact that stem and progenitor cells exist in adulthood rises the question where are they located and how are they mobilized. A model of stem cell niches and the contribution of the vasculature to these structures was proposed by Moore and Lemischka (2006; Bautch 2011). The concept proposes the existence of several organ specific stem and progenitor cells residing in the respective areas of the vasculature. One of the best described is the hematopoietic bone marrow niche for HSCs. This consists of the osteogenic niche which maintains HSCs in a quiescent state, whereas the connected vascular niche regulates the differentiation and mobilization of progenitor cells (Bianco 2011; Dimmeler 2010). Furthermore, in the central nervous system neural stem cells reside closely to blood vessels at two sites: the subventricular zone and the subgranular zone (Goldberg and Hirschi 2009). Stem cell self renewal was shown to occur under participation of Notch receptors and Hes1 as target gene in coculture experiments with endothelial cells (Shen et al. 2004).

Additional studies in humans identified progenitor cells in vessels like human adult vena saphena or zones localized between smooth muscle and adventitial layer of the human adult vascular wall (Campagnolo et al. 2010; Zengin et al. 2006). A specific niche in the vasculature seems to be constituted by the close proximity of endothelial cells with pericytes, where these display some properties of stem and progenitor cells when they detach from the endothelial cells (Bianco 2011; Bianco et al. 2008), but further studies are needed to clarify the role and importance of this aspect.

Endothelial Progenitor Cells in Therapy

EPCs of various kind have been extensively applied in different experimental settings to stimulate angiogenesis and have been demonstrated to be beneficial for vascular repair (Fadini et al. 2010; Lipinski et al. 2007; Dimmeler et al. 2008). But comparative studies have been hampered by the fact, that no consistent definition of the EPCs existed and every study used different methods for isolation and characterization of their cell population. In a most recent review Fadini et al. (2012) present an elaborate overview over the diverse nomenclature of EPCs and the different approaches to isolate and characterize the EPCs in regard to vessel inducing capacity.

The different cell types studied showed either beneficial effects in neovascularization, presumably via paracrine factors, or as in the case of ECFCs (also termed late outgrowth endothelial cells, BOECs) were demonstrated to also incorporate directly into newly formed vessels.

Cells obtained from short term culture conditions (early EPCs, CFU-ECs) are suggested to support angiogenesis only by secreting paracrine factors (Rehman et al. 2003; Urbich et al. 2005), they are found associated with newly formed vessels, support postischemic angiogenesis, but experiments failed to show long term integration such as in a mouse myocardial infarction model (Ziebart et al. 2008). For example, Kawamoto et al. (2003) investigated transplantation of autologous EPCs obtained from $CD31^{+}$MNCs in a chronic rat myocardial ischemia model and showed a favourable outcome in this setting. Similarly, G-SCF mobilized human autologous bone marrow-derived progenitor cells implanted for revascularization reduced morbidity and mortality due to left ventricular remodeling after myocardial infarction in a rat model (Kocher et al. 2001; Itescu et al. 2003). Furthermore, *ex vivo* expanded hematopoietic stem cells from cord blood which were selected for $CD133^{+}$ and intravenously injected displayed migration to the ischemic area and improvement of myocardial function (Schlechta et al. 2010).

Still a major question remaining is that it has not been rigorously proven what is the physiological mode of action of circulating EPCs in an *in vivo* setting, especially corresponding data from mouse models are largely missing. It is also under debate whether the different progenitor populations that can be gained under specific conditions in cell culture indeed occur *in vivo* in the bloodstream. Furthermore, the more EPCs are investigated, the larger appears the complexity because the plasticity of these cells often obscures precise classification. It has also to be considered that keeping cells in culture might change their program and that cells from an artificial environment might not exist in the same form *in vivo*.

Nevertheless, a most attractive hypothesis would be that at least some EPCs, such as ECFCs/BOECs in humans, are shed from foci of progenitor cells existing in the vessel wall of capillaries and function to replace aged endothelial cells throughout the body and to facilitate neovascularization of ischemic areas following their increased mobilization upon wounding or other injuries. There is at least some recent evidence also from the murine model in favor of the existence of such progenitor cells in the vessel wall of microvessels (Naito et al. 2011).

Cells originating after long term culture from some not well characterized early progenitors are a very rare, hidden population (termed late EPCs, ECFCs or BOECs). These cells are distinct as they physically incorporate into capillaries (Hur et al. 2004; Richardson and Yoder 2011; Yoder et al. 2007). Importantly, a large-scale animal protein-free humanized expansion strategy was described by Reinisch et al. (2009) which provides a basis for a therapeutic use of ECFCs.

The beneficial effect of human BOECs on wound healing appears to be twofold, it is mediated by growth factor cross-talk as well as direct cell integration supporting new vascularisation. These cells have for example been used in a combined tissue engineering approach for repairing large skin defects (Hendrickx et al. 2010). A further exciting finding was, that using BOECs for infusion after acute myocardial infarction in a porcine model leads to enhanced neovascularisation, some stable integration of the cells into neovessels and improves transfusion and heart function, although it does not mediate cardiomyogenesis (Dubois et al. 2010). However, as improved transfusion is the required first step in the repair of any ischemic tissue, these findings suggest that combined infusion of ECFCs/BOECs with other modalities such as growth factors triggering cardiomyogenesis might result in synergistic effects.

In regard of generating ECFCs/BOECs, both sources, peripheral or cord blood, are currently explored and have specific advantages: stem/progenitor cells from peripheral blood would be easily obtainable for autologous application from the patient and therefore carry minimal risk of unwanted side effects; on the other hand, cells from cord blood are available from numerous recently established huge cord blood collections, they can be obtained HLA-matched and display superior proliferative capacity which makes them especially useful for triggering of neovascularization in ischemic diseases and for engineering of prevascularized tissue patches, e.g. of heart muscle, for regenerative purposes.

With a different goal in mind, we have previously shown that tumor targeting by *ex vivo* expanded ECFCs is possible and could potentially be used in tumor therapy to carry inhibitory factors to the tumor site. When we transduced ECFCs with an adenovirus to secrete a soluble form of colony stimulating factor-1 receptor (CD115) that inhibits macrophage viability and migration, and injected these cells into a prostate tumor xenograft model, homing of the cells to regions of ischemia in the growing tumors was observed. Furthermore, the soluble CD115 secreted by the cells obviously blocked the activation of macrophages in the tumor environment and thus suppressed the growth of the prostate tumor (Lucas et al. 2009).

In another setting, EPCs have also been proposed to be biomarkers for cardiovascular disease (Sen et al. 2011). Since EPCs are not a defined population with a specific marker it is reasonable to characterize them by one stem cell marker and one endothelial marker. For historical reasons frequently $CD34^{+}KDR^{+}$ is used. However, also hematopoietic stem/progenitor cells display these markers and it is therefore unclear which kind of stem/progenitor cells were scored in those experiments. It was suggested to use CD133 in addition, which is only found on early EPCs, but then the number of positive cells decreases and makes analysis more

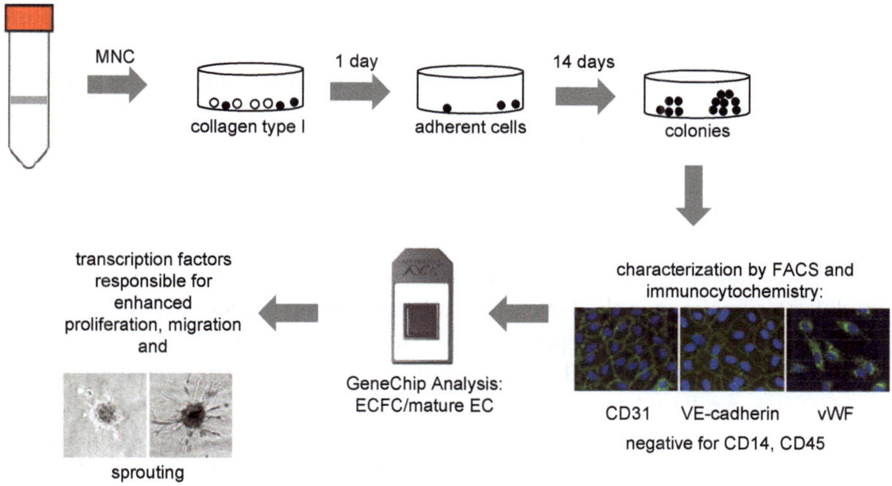

Fig. 3.1 Isolation procedure and characterization of ECFCs

difficult. It is therefore a complex issue to specify the markers for using these cells as clinical biomarkers due to different cell types displaying the same surface proteins.

Characterization of ECFCs in Comparison to Mature Endothelial Cells and Other Stem/Progenitor Cells

Based on the encouraging data by Dubois et al. (2010) using ECFCs/BOECs in a porcine myocardial infarct model and the lack of data describing the unambiguous and consistent definition of markers for these progenitor cells, we analyzed the gene expression profile of ECFCs of umbilical cord blood and adult peripheral blood in comparison to mature vessel wall endothelial cells and $CD34^+$ stem cells (Fig. 3.1).

Mononuclear cells were isolated from the blood by Ficoll density gradient centrifugation and cultured on collagen I coated plates for 10–14 days as described by Yoder et al. (2007). After propagation, the absence of the hematopoietic markers CD14 and CD45 and the presence of endothelial surface markers such as CD34, KDR, CD31, VE-cadherin and vWF was determined. Generally, the expression of the endothelial markers on ECFCs was comparable to vessel wall endothelial cells (human umbilical vein endothelial cells, HUVEC) in FACS analysis and immunocytochemistry. However, when we performed gene profiling comparing cord blood ECFCs and HUVEC or ECFCs from adult peripheral blood and adult venous endothelial cells, distinct differences for a small number of genes were observed. These probably constitute genes responsible for the progenitor status of the ECFCs

(Fig. 3.1). For control purposes we also analyzed CD34$^+$ stem cells isolated by positive magnetic separation from the mononuclear fraction.

Based on these results, we defined transcription factors and secreted proteins expressed at higher extent in ECFCs than in HUVECs. Some of these factors were not expressed in the largely hematopoietic CD34$^+$ stem cells, however others were. The most strongly and differentially expressed transcription factor was FOXF1, a member of the large family of forkhead box transcription factors which are known as key regulators of differentiation to different cell lineages. Using FOXF1 in gain and loss-of-function studies in ECFCs and HUVEC we delineated that the factor modulates signaling via the Notch-Dll pathway leading to increased sprouting activity *in vitro* (Lipnik et al. in preparation). The obtained data indicate a crucial role for FOXF1 for the endothelial progenitor status and further support an involvement of ECFCs in capillary formation and arterial specification.

Considering the stimulatory function of ECFCs for new vessel formation *in vivo* we further investigated if they also specifically produce factors which would be beneficial for the initiation of vascularization. From the gene profiling studies comparing ECFCs with vessel wall endothelial cells we also identified a secreted factor with high expression in ECFCs which stimulates proliferation, migration and sprouting of ECFCs as well as mature endothelial cells (Lipnik et al. in preparation). When secreted by ECFCs it could therefore work in an autocrine and paracrine fashion and provide increased proliferative, migratory and sprouting capacities to the progenitors after homing into ischemic vessel areas, as well as to neighbouring mature endothelial cells and thereby trigger vascular sprouting. This factor is currently evaluated in rodent models of myocardial infarction for its potential activity to support revascularization and heart regeneration.

Conclusions

Attempts to define the identity and the precise nature of various cells termed EPCs is hurdled by their described apparent variety and plasticity. Additional investigations using more sophisticated techniques of life imaging and lineage tracing will be needed to gain better knowledge of these various cells. At the current stage, it seems possible that endothelial progenitor cells exist in various forms and are of different origin, they appear to be dynamic and to respond to environmental changes. However, taken together, currently available data suggest that the cells which collectively can be described as early outgrowth cells and display hematopoietic markers, can stimulate neovascularization by paracrine activities, but do not incorporate into the endothelial layer of neovessels. In contrast, the fraction of cells described as late outgrowth cells induce vascularization and are found at least in part integrated into the newly formed vessels. The latter may therefore be true endothelial progenitors and it has been suggested that they are not of hematopoietic origin, but may bud from the vessel wall and may function as vascular specific progenitor cells. Whether these cells fulfill indeed this function *in vivo* remains to be shown.

However, irrespective of their physiological role and due to the easiness of their *ex vivo* production, these cells are of high interest for cell therapies, e.g. to increase perfusion in all kinds of ischemic diseases and even for the targeting of tumors. We propose that gaining better understanding of this cell type will not only bring new insights how the vascular system regenerates itself, but will also lead to new developments in all kinds of therapeutic tissue regeneration and could reveal diagnostic features for cardiovascular diseases. Initial progress in these directions was recently achieved by us by the characterization of factors contributing to the progenitor status of ECFC/BOEC and mediating increased neovascularization capacity.

Acknowledgements This work was supported by grants of the European Commission (Health-F2-2009-222995, INELPY/Infarct Cell Therapy), the Austrian Science Fund (FWF P-21291-B11) and the Research Studio Austria Programme of the FFG (FFG/RSA 832042).

References

Anghelina M, Krishnan P, Moldovan L, Moldovan NI (2004) Monocytes and macrophages form branched cell columns in matrigel: implications for a role in neovascularization. Stem Cells Dev 13(6):665–676

Anghelina M, Krishnan P, Moldovan L, Moldovan NI (2006) Monocytes/macrophages cooperate with progenitor cells during neovascularization and tissue repair: conversion of cell columns into fibrovascular bundles. Am J Pathol 168(2):529–541

Arciniegas E, Frid MG, Douglas IS, Stenmark KR (2007) Perspectives on endothelial-to-mesenchymal transition: potential contribution to vascular remodeling in chronic pulmonary hypertension. Am J Physiol 293(1):L1–L8

Asahara T, Murohara T, Sullivan A, Silver M, van der Zee R, Li T, Witzenbichler B, Schatteman G, Isner JM (1997) Isolation of putative progenitor endothelial cells for angiogenesis. Science 275(5302):964–967

Au P, Tam J, Fukumura D, Jain RK (2008) Bone marrow-derived mesenchymal stem cells facilitate engineering of long-lasting functional vasculature. Blood 111(9):4551–4558

Bautch VL (2011) Stem cells and the vasculature. Nat Med 17(11):1437–1443

Bertrand JY, Giroux S, Golub R, Klaine M, Jalil A, Boucontet L, Godin I, Cumano A (2005) Characterization of purified intraembryonic hematopoietic stem cells as a tool to define their site of origin. Proc Nat Acad Sci U S A 102(1):134–139

Bertrand JY, Chi NC, Santoso B, Teng S, Stainier DY, Traver D (2010) Haematopoietic stem cells derive directly from aortic endothelium during development. Nature 464(7285):108–111

Bianco P (2011) Bone and the hematopoietic niche: a tale of two stem cells. Blood 117(20):5281–5288

Bianco P, Robey PG, Simmons PJ (2008) Mesenchymal stem cells: revisiting history, concepts, and assays. Cell Stem Cell 2(4):313–319

Boisset JC, van Cappellen W, Andrieu-Soler C, Galjart N, Dzierzak E, Robin C (2010) In vivo imaging of haematopoietic cells emerging from the mouse aortic endothelium. Nature 464(7285):116–120

Campagnolo P, Cesselli D, Al Haj Zen A, Beltrami AP, Krankel N, Katare R, Angelini G, Emanueli C, Madeddu P (2010) Human adult vena saphena contains perivascular progenitor cells endowed with clonogenic and proangiogenic potential. Circulation 121(15):1735–1745

Cogle CR, Wainman DA, Jorgensen ML, Guthrie SM, Mames RN, Scott EW (2004) Adult human hematopoietic cells provide functional hemangioblast activity. Blood 103(1):133–135

Dimmeler S (2010) Regulation of bone marrow-derived vascular progenitor cell mobilization and maintenance. Arterioscler Thromb Vasc Biol 30(6):1088–1093

Dimmeler S, Burchfield J, Zeiher AM (2008) Cell-based therapy of myocardial infarction. Arterioscler Thromb Vasc Biol 28(2):208–216
Dubois C, Liu X, Claus P, Marsboom G, Pokreisz P, Vandenwijngaert S, Depelteau H, Streb W, Chaothawee L, Maes F, Gheysens O, Debyser Z, Gillijns H, Pellens M, Vandendriessche T, Chuah M, Collen D, Verbeken E, Belmans A, Van de Werf F, Bogaert J, Janssens S (2010) Differential effects of progenitor cell populations on left ventricular remodeling and myocardial neovascularization after myocardial infarction. J Am Coll Cardiol 55(20):2232–2243
Eilken HM, Nishikawa S, Schroeder T (2009) Continuous single-cell imaging of blood generation from haemogenic endothelium. Nature 457(7231):896–900
Fadini GP, Agostini C, Avogaro A (2010) Autologous stem cell therapy for peripheral arterial disease meta-analysis and systematic review of the literature. Atherosclerosis 209(1):10–17
Fadini GP, Losordo D, Dimmeler S (2012) Critical reevaluation of endothelial progenitor cell phenotypes for therapeutic and diagnostic use. Circ Res 110(4):624–637
Ferkowicz MJ, Yoder MC (2005) Blood island formation: longstanding observations and modern interpretations. Exp Hematol 33(9):1041–1047
Folkman J, Shing Y (1992) Angiogenesis. J Biol Chem 267(16):10931–10934
Friedrich EB, Walenta K, Scharlau J, Nickenig G, Werner N (2006) CD34-/CD133+/VEGFR-2+ endothelial progenitor cell subpopulation with potent vasoregenerative capacities. Circ Res 98(3):e20–e25
Gao D, Nolan DJ, Mellick AS, Bambino K, McDonnell K, Mittal V (2008) Endothelial progenitor cells control the angiogenic switch in mouse lung metastasis. Science 319(5860):195–198
Gehling UM, Ergun S, Schumacher U, Wagener C, Pantel K, Otte M, Schuch G, Schafhausen P, Mende T, Kilic N, Kluge K, Schafer B, Hossfeld DK, Fiedler W (2000) In vitro differentiation of endothelial cells from AC133-positive progenitor cells. Blood 95(10):3106–3112
Gehling UM, Ergun S, Fiedler W (2007) CFU-EC: how they were originally defined. Blood 110(3):1073
Goldberg JS, Hirschi KK (2009) Diverse roles of the vasculature within the neural stem cell niche. Regen Med 4(6):879–897
Gothert JR, Gustin SE, van Eekelen JA, Schmidt U, Hall MA, Jane SM, Green AR, Gottgens B, Izon DJ, Begley CG (2004) Genetically tagging endothelial cells in vivo: bone marrow-derived cells do not contribute to tumor endothelium. Blood 104(6):1769–1777
Gulati R, Jevremovic D, Peterson TE, Chatterjee S, Shah V, Vile RG, Simari RD (2003) Diverse origin and function of cells with endothelial phenotype obtained from adult human blood. Circ Res 93(11):1023–1025
Gunsilius E, Duba HC, Petzer AL, Kahler CM, Grunewald K, Stockhammer G, Gabl C, Dirnhofer S, Clausen J, Gastl G (2000) Evidence from a leukaemia model for maintenance of vascular endothelium by bone-marrow-derived endothelial cells. The Lancet 355(9216):1688–1691
Haar JL, Ackerman GA (1971) A phase and electron microscopic study of vasculogenesis and erythropoiesis in the yolk sac of the mouse. Anat Rec 170(2):199–223
Hendrickx B, Verdonck K, Van den Berge S, Dickens S, Eriksson E, Vranckx JJ, Luttun A (2010) Integration of blood outgrowth endothelial cells in dermal fibroblast sheets promotes full thickness wound healing. Stem Cells 28(7):1165–1177
Hur J, Yoon CH, Kim HS, Choi JH, Kang HJ, Hwang KK, Oh BH, Lee MM, Park YB (2004) Characterization of two types of endothelial progenitor cells and their different contributions to neovasculogenesis. Arterioscler Thromb Vasc Biol 24(2):288–293
Hur J, Yang HM, Yoon CH, Lee CS, Park KW, Kim JH, Kim TY, Kim JY, Kang HJ, Chae IH, Oh BH, Park YB, Kim HS (2007) Identification of a novel role of T cells in postnatal vasculogenesis: characterization of endothelial progenitor cell colonies. Circulation 116(15):1671–1682
Ingram DA, Mead LE, Tanaka H, Meade V, Fenoglio A, Mortell K, Pollok K, Ferkowicz MJ, Gilley D, Yoder MC (2004) Identification of a novel hierarchy of endothelial progenitor cells using human peripheral and umbilical cord blood. Blood 104(9):2752–2760
Ingram DA, Caplice NM, Yoder MC (2005a) Unresolved questions, changing definitions, and novel paradigms for defining endothelial progenitor cells. Blood 106(5):1525–1531

Ingram DA, Mead LE, Moore DB, Woodard W, Fenoglio A, Yoder MC (2005b) Vessel wall-derived endothelial cells rapidly proliferate because they contain a complete hierarchy of endothelial progenitor cells. Blood 105(7):2783–2786

Ishikawa M, Asahara T (2004) Endothelial progenitor cell culture for vascular regeneration. Stem Cells Dev 13(4):344–349

Itescu S, Kocher AA, Schuster MD (2003) Myocardial neovascularization by adult bone marrow-derived angioblasts: strategies for improvement of cardiomyocyte function. Heart Fail Rev 8(3):253–258. doi:5138694 [pii]

Jujo K, Ii M, Losordo DW (2008) Endothelial progenitor cells in neovascularization of infarcted myocardium. J Mol Cell Cardiol 45(4):530–544

Kawamoto A, Tkebuchava T, Yamaguchi J, Nishimura H, Yoon YS, Milliken C, Uchida S, Masuo O, Iwaguro H, Ma H, Hanley A, Silver M, Kearney M, Losordo DW, Isner JM, Asahara T (2003) Intramyocardial transplantation of autologous endothelial progenitor cells for therapeutic neovascularization of myocardial ischemia. Circulation 107(3):461–468

Kerbel RS, Benezra R, Lyden DC, Hattori K, Heissig B, Nolan DJ, Mittal V, Shaked Y, Dias S, Bertolini F, Rafii S (2008) Endothelial progenitor cells are cellular hubs essential for neoangiogenesis of certain aggressive adenocarcinomas and metastatic transition but not adenomas. Proc Nat Acad Sci U S A 105(34):E54; author reply E55

Kissa K, Herbomel P (2010) Blood stem cells emerge from aortic endothelium by a novel type of cell transition. Nature 464(7285):112–115

Kocher AA, Schuster MD, Szabolcs MJ, Takuma S, Burkhoff D, Wang J, Homma S, Edwards NM, Itescu S (2001) Neovascularization of ischemic myocardium by human bone-marrow-derived angioblasts prevents cardiomyocyte apoptosis, reduces remodeling and improves cardiac function. Nat Med 7(4):430–436

Kovacic JC, Moore J, Herbert A, Ma D, Boehm M, Graham RM (2008) Endothelial progenitor cells, angioblasts, and angiogenesis–old terms reconsidered from a current perspective. Trends Cardiovasc Med 18(2):45–51

Lancrin C, Sroczynska P, Stephenson C, Allen T, Kouskoff V, Lacaud G (2009) The haemangioblast generates haematopoietic cells through a haemogenic endothelium stage. Nature 457(7231):892–895

Lin Y, Weisdorf DJ, Solovey A, Hebbel RP (2000) Origins of circulating endothelial cells and endothelial outgrowth from blood. J Clin Invest 105(1):71–77

Lipinski MJ, Biondi-Zoccai GG, Abbate A, Khianey R, Sheiban I, Bartunek J, Vanderheyden M, Kim HS, Kang HJ, Strauer BE, Vetrovec GW (2007) Impact of intracoronary cell therapy on left ventricular function in the setting of acute myocardial infarction: a collaborative systematic review and meta-analysis of controlled clinical trials. J Am Coll Cardiol 50(18):1761–1767

Lucas T, Abraham D, Untergasser G, Zins K, Hofer E, Gunsilius E, Aharinejad S (2009) Adenoviral-mediated endothelial precursor cell delivery of soluble CD115 suppresses human prostate cancer xenograft growth in mice. Stem Cells 27(9):2342–2352

Moldovan NI, Goldschmidt-Clermont PJ, Parker-Thornburg J, Shapiro SD, Kolattukudy PE (2000) Contribution of monocytes/macrophages to compensatory neovascularization: the drilling of metalloelastase-positive tunnels in ischemic myocardium. Circ Res 87(5):378–384

Moore KA, Lemischka IR (2006) Stem cells and their niches. Science 311(5769):1880–1885

Murohara T, Ikeda H, Duan J, Shintani S, Sasaki K, Eguchi H, Onitsuka I, Matsui K, Imaizumi T (2000) Transplanted cord blood-derived endothelial precursor cells augment postnatal neovascularization. J Clin Invest 105(11):1527–1536

Naito H, Kidoya H, Sakimoto S, Wakabayashi T, Takakura N (2011) Identification and characterization of a resident vascular stem/progenitor cell population in preexisting blood vessels. EMBO J 31(4):842–855

Nolan DJ, Ciarrocchi A, Mellick AS, Jaggi JS, Bambino K, Gupta S, Heikamp E, McDevitt MR, Scheinberg DA, Benezra R, Mittal V (2007) Bone marrow-derived endothelial progenitor cells are a major determinant of nascent tumor neovascularization. Genes Dev 21(12):1546–1558

Park C, Ma YD, Choi K (2005) Evidence for the hemangioblast. Exp Hematol 33(9):965–970

Peichev M, Naiyer AJ, Pereira D, Zhu Z, Lane WJ, Williams M, Oz MC, Hicklin DJ, Witte L, Moore MA, Rafii S (2000) Expression of VEGFR-2 and AC133 by circulating human CD34(+) cells identifies a population of functional endothelial precursors. Blood 95(3):952–958

Pelosi E, Valtieri M, Coppola S, Botta R, Gabbianelli M, Lulli V, Marziali G, Masella B, Muller R, Sgadari C, Testa U, Bonanno G, Peschle C (2002) Identification of the hemangioblast in postnatal life. Blood 100(9):3203–3208

Popa ER, Harmsen MC, Tio RA, van der Strate BW, Brouwer LA, Schipper M, Koerts J, De Jongste MJ, Hazenberg A, Hendriks M, van Luyn MJ (2006) Circulating CD34 + progenitor cells modulate host angiogenesis and inflammation in vivo. J Mol Cell Cardiol 41(1):86–96

Potenta S, Zeisberg E, Kalluri R (2008) The role of endothelial-to-mesenchymal transition in cancer progression. Br J Cancer 99(9):1375–1379

Purhonen S, Palm J, Rossi D, Kaskenpaa N, Rajantie I, Yla-Herttuala S, Alitalo K, Weissman IL, Salven P (2008) Bone marrow-derived circulating endothelial precursors do not contribute to vascular endothelium and are not needed for tumor growth. Proc Nat Acad Sci U S A 105(18):6620–6625

Rehman J, Li J, Orschell CM, March KL (2003) Peripheral blood "endothelial progenitor cells" are derived from monocyte/macrophages and secrete angiogenic growth factors. Circulation 107(8):1164–1169

Reinisch A, Hofmann NA, Obenauf AC, Kashofer K, Rohde E, Schallmoser K, Flicker K, Lanzer G, Linkesch W, Speicher MR, Strunk D (2009) Humanized large-scale expanded endothelial colony-forming cells function in vitro and in vivo. Blood 113(26):6716–6725

Richardson MR, Yoder MC (2011) Endothelial progenitor cells: quo vadis? J Mol Cell Cardiol 50(2):266–272

Risau W, Sariola H, Zerwes HG, Sasse J, Ekblom P, Kemler R, Doetschman T (1988) Vasculogenesis and angiogenesis in embryonic-stem-cell-derived embryoid bodies. Development 102(3):471–478

Rohde E, Malischnik C, Thaler D, Maierhofer T, Linkesch W, Lanzer G, Guelly C, Strunk D (2006) Blood monocytes mimic endothelial progenitor cells. Stem Cells 24(2):357–367

Rohde E, Bartmann C, Schallmoser K, Reinisch A, Lanzer G, Linkesch W, Guelly C, Strunk D (2007) Immune cells mimic the morphology of endothelial progenitor colonies in vitro. Stem Cells 25(7):1746–1752

Romagnani P, Annunziato F, Liotta F, Lazzeri E, Mazzinghi B, Frosali F, Cosmi L, Maggi L, Lasagni L, Scheffold A, Kruger M, Dimmeler S, Marra F, Gensini G, Maggi E, Romagnani S (2005) CD14+CD34low cells with stem cell phenotypic and functional features are the major source of circulating endothelial progenitors. Circ Res 97(4):314–322. doi:10.1161/01.RES.0000177670.72216.9b [pii]

Sasaki T, Mizuochi C, Horio Y, Nakao K, Akashi K, Sugiyama D (2010) Regulation of hematopoietic cell clusters in the placental niche through SCF/Kit signaling in embryonic mouse. Development 137(23):3941–3952

Schlechta B, Wiedemann D, Kittinger C, Jandrositz A, Bonaros NE, Huber JC, Preisegger KH, Kocher AA (2010) Ex-vivo expanded umbilical cord blood stem cells retain capacity for myocardial regeneration. Circ J 74(1):188–194

Schwartz SM, Benditt EP (1976) Clustering of replicating cells in aortic endothelium. Proc Nat Acad Sci U S A 73(2):651–653

Sen S, McDonald SP, Coates PT, Bonder CS (2011) Endothelial progenitor cells: novel biomarker and promising cell therapy for cardiovascular disease. Clin Sci 120(7):263–283

Shen Q, Goderie SK, Jin L, Karanth N, Sun Y, Abramova N, Vincent P, Pumiglia K, Temple S (2004) Endothelial cells stimulate self-renewal and expand neurogenesis of neural stem cells. Science 304(5675):1338–1340

Strijbos MH, Rao C, Schmitz PI, Kraan J, Lamers CH, Sleijfer S, Terstappen LW, Gratama JW (2008) Correlation between circulating endothelial cell counts and plasma thrombomodulin levels as markers for endothelial damage. Thromb Haemost 100(4):642–647

Szabo E, Rampalli S, Risueno RM, Schnerch A, Mitchell R, Fiebig-Comyn A, Levadoux-Martin M, Bhatia M (2010) Direct conversion of human fibroblasts to multilineage blood progenitors. Nature 468(7323):521–526

Urbich C, Aicher A, Heeschen C, Dernbach E, Hofmann WK, Zeiher AM, Dimmeler S (2005) Soluble factors released by endothelial progenitor cells promote migration of endothelial cells and cardiac resident progenitor cells. J Mol Cell Cardiol 39(5):733–742

Yoder MC, Mead LE, Prater D, Krier TR, Mroueh KN, Li F, Krasich R, Temm CJ, Prchal JT, Ingram DA (2007) Redefining endothelial progenitor cells via clonal analysis and hematopoietic stem/progenitor cell principals. Blood 109(5):1801–1809

Yoder MC (2010) Is endothelium the origin of endothelial progenitor cells? Arterioscler Thrombosis Vasc Biol 30(6):1094–1103

Yoon CH, Hur J, Park KW, Kim JH, Lee CS, Oh IY, Kim TY, Cho HJ, Kang HJ, Chae IH, Yang HK, Oh BH, Park YB, Kim HS (2005) Synergistic neovascularization by mixed transplantation of early endothelial progenitor cells and late outgrowth endothelial cells: the role of angiogenic cytokines and matrix metalloproteinases. Circulation 112(11):1618–1627

Zape JP, Zovein AC (2011) Hemogenic endothelium: origins, regulation, and implications for vascular biology. Semin Cell Dev Biol 22(9):1036–1047

Zengin E, Chalajour F, Gehling UM, Ito WD, Treede H, Lauke H, Weil J, Reichenspurner H, Kilic N, Ergun S (2006) Vascular wall resident progenitor cells: a source for postnatal vasculogenesis. Development 133(8):1543–1551

Ziebart T, Yoon CH, Trepels T, Wietelmann A, Braun T, Kiessling F, Stein S, Grez M, Ihling C, Muhly-Reinholz M, Carmona G, Urbich C, Zeiher AM, Dimmeler S (2008) Sustained persistence of transplanted proangiogenic cells contributes to neovascularization and cardiac function after ischemia. Circ Res 103(11):1327–1334

Zovein AC, Hofmann JJ, Lynch M, French WJ, Turlo KA, Yang Y, Becker MS, Zanetta L, Dejana E, Gasson JC, Tallquist MD, Iruela-Arispe ML (2008) Fate tracing reveals the endothelial origin of hematopoietic stem cells. Cell Stem Cell 3(6):625–636

Chapter 4
Adipose-derived Stromal/Stem Cells and Their Differentiation Potential into the Endothelial Lineage

Patrick C. Baer and Werner Luttmann

Abstract Subcutaneous adipose tissue is ubiquitous and easily accessible in large quantities with a minimal invasive procedure by liposuction aspiration, qualifying it as a good source for stem cell isolation. Adipose-derived mesenchymal stromal/stem cells (ASCs) can easily be isolated by a relatively simple method, yielding a high amount of stem cells, which is essential for stem cell-based therapies and tissue engineering. Several studies have provided evidence that ASCs in situ reside in the wall surrounding the vasculature, and that blood vessels in virtually all organs and tissues harbour ubiquitous (mesenchymal) stem cells in their perivascular niche. Nevertheless, the exact localization of ASCs in native adipose tissue is still under debate. ASCs are isolated by their capacity to adhere to cell culture plastic and are maintained in culture for up to 10–15 passages. Cultured cells are characterized by their expression of a panel of characteristic markers and their capacity to differentiate into cells from meso-, ecto- and endodermal lineages. ASCs possess a high plasticity and differentiate into various cell types, including endothelial cells *in vitro* and *in vivo*. This review summarizes and discusses the current knowledge about ASCs and their differentiation potential into endothelial cells *in vitro* and *in vivo*.

Introduction

Research on stem cell sources for cell therapeutical interventions for tissue regeneration support has accelerated during the past two decades. Ideal stem cells for use in therapeutical approaches need to be isolated with minimal harm for the patient, must be available in high cell numbers, proliferate in culture, and differentiate into a broad spectrum of lineages. Adipose-derived mesenchymal stromal/stem cells (ASCs) fulfill these requirements: they can be isolated by a minimally invasive harvesting procedure, achieve a high cell yield with a large proliferative activity, and

P. C. Baer (✉)
Division of Nephrology, Department of Internal Medicine III,
Johann Wolfgang Goethe University, Theodor-Stern-Kai 7,
60590 Frankfurt am Main, Germany
e-mail: p.baer@em.uni-frankfurt.de

W. Luttmann
EuroBioSciences GmbH, Friesoythe, Germany

J. Hescheler, E. Hofer (eds.), *Adult and Pluripotent Stem Cells*,
DOI 10.1007/978-94-017-8657-7_4, © Springer Science+Business Media Dordrecht 2014

possess a remarkable plasticity. They are able to differentiate into cells from meso-, ecto- and endodermal origin *in vitro* and *in vivo* (e.g. adipocytes, osteoblasts, chondrocytes, myocytes, and neural, epithelial and endothelial cells (Baer 2011; Brzoska et al. 2005; Cao et al. 2005; Fraser et al. 2006; Gimble and Guilak 2003a, b; Lee and Kemp 2006; Miranville et al. 2004; Mizuno et al. 2002; Planat-Benard et al. 2004; Rodriguez et al. 2006; Strem et al. 2005; Zuk et al. 2002, 2001). Furthermore, the use of ASCs has been proven to be effective in clinical studies over the past years. The use of ASCs from an autologous origin seems to be a safe source for cell-based regenerative approaches. In addition, ASCs have reduced immunogenic properties and an immunosuppressive potential, which make them also attractive for allogenic stem cell therapy (da Silva Meirelles et al. 2006; Kassis et al. 2006; Zou et al. 2010).

In general, mesenchymal stromal/stem cells (MSCs) can be isolated from almost every tissue of the body, including bone marrow, adipose tissue, peripheral blood, menstrual blood, solid organs (e.g. liver, spleen, thymus, lung) and many other adult tissues (e.g. synovium, dermis, periosteum, deciduous teeth), as well as umbilical cord blood and fetal tissues (Hoogduijn et al. 2011; McIntosh et al. 2006; McIntosh 2011). MSCs are a rare and quiescent population in their niche within fully specialized tissues. At present, there is a strong amount of data indicating that MSCs represent independent population(s) of stem cells with self-renewal properties and established multipotent differentiation profile *in vitro* (Dominici et al. 2009). Nevertheless, MSC preparations are heterogeneous cell cultures comprising a subset of stem cells (or different subsets of stem cells) and more differentiated (progenitor) cells. In general, MSCs are isolated by their capacity to adhere to culture-dish plastic. The cells can be expanded in culture while maintaining their differentiation potential during standard cell culture, and are characterized by a specific panel of markers. However, the characterization of MSCs remains difficult due to the lack of a definitive and unique cellular marker. Therefore, the International Society for Cellular Therapy proposed three minimal criteria for the definition of MSCs: (a) Plastic-adherence, (b) expression of CD73, CD90, and CD105, and lack of CD11b or CD14, CD19 or CD79α, CD45, and HLA-DR expression, and (c) their tri-lineage differentiation potential into adipocytes, chondrocytes and osteoblasts (Dominici et al. 2006).

ASCs: Tissue Localization, Isolation and Culture

Several studies have tried to identify ASCs (and their subpopulations) within intact adipose tissue. This is a complicated endeavour because no single marker specifically and unequivocally identifies undifferentiated ASCs. Histological studies using immunohistochemical and immunofluorescence techniques suggest that a stem cell population (or populations) resides in a perivascular location, where ASCs coexist with pericytes and endothelial cells (Cai et al. 2011; Corselli et al. 2010; Crisan et al. 2009, 2008; Fraser et al. 2008; Lin et al. 2010, 2008; Traktuev et al. 2009; Zannettino et al. 2008; Zimmerlin et al. 2010). It has also been suggested that ASCs (and MSCs in general) are a subset of pericytes or vascular stem/precursor cells at various stages of differentiation located in the wall surrounding the vasculature (Lin et al. 2008), and that blood vessels in virtually all organs and tissues harbour ubiquitous (mesenchymal) stem cells in their perivascular niche (Corselli et al. 2010). Whereas all these studies provided much evidence and it seems likely that ASCs in situ reside in a perivascular niche in a $CD34^{+}/CD90^{+}/CD31^{-}/CD45^{-}/CD146^{-}$ phenotype, the definite identification of the ASC population(s) in situ has currently not been achieved. The data currently available are not adequate for the clear definition of an exclusive ASC population in situ.

Subcutaneous adipose tissue is ubiquitous and easily accessible in large quantities with a minimal invasive procedure by liposuction aspiration. Liposuction surgery is a well-tolerated and safe procedure yielding large quantities of aspirate, qualifying as a good source of (autologous) stem cells for further cell isolation. ASCs can easily be isolated by a relatively simple method from liposuction aspirates (as well as from adipose tissue biopsies or dermolipectomy). Briefly, the liposuction aspirate (or the biopsie tissue) is digested with collagenase and centrifuged in several steps, followed by the outgrowth of the plastic adherent fraction from the primary isolated cell mixture (Fig. 4.1; Zuk et al. 2002, 2001). It has been described that stem and progenitor cells in the primary isolated, uncultured fraction from adipose tissue (stroma-vascular fraction, SVF) usually amount to up to 3 %, and this is approximately 2,500-fold more than the frequency of stem cells in bone marrow (Fraser et al. 2008). The number of cells that can be isolated from liposuction aspirates is approximately $0.5-2.0\times10^{6}$ cells/g of adipose tissue (Aust et al. 2004; De Ugarte et al. 2003; Oedayrajsingh-Varma et al. 2006; Zhu et al. 2008), whereby the frequency of (adherent) stromal/stem cells differs in many studies (range from 1 to 10 %), most likely depending on the donor and tissue harvesting site and/or the isolation method.

Adherent cells cultured in standard culture medium are considered as multipotent ASCs (Fig. 4.1). The cells can be cultured by serial passaging without losing their differentiation potential (Aust et al. 2004; De Ugarte et al. 2003; Oedayrajsingh-Varma et al. 2006; Zhu et al. 2008) and have the capacity to maintain chromosome stability in long-term cultures (Grimes et al. 2009). Early during primary culture ASCs diplay a slightly heterogeneous morphology indicating different stem and precursor cell subpopulations, and (maybe) more differentiated cells (dedifferentiated endothelial cells, smooth muscle cells and pericytes). Nevertheless, it has been demonstrated that more than 85 % of the initially adhered cells had a $CD31^{-}/CD34^{+}/CD45^{-}/CD146^{-}$ phenotype, and only few endothelial cells, leucocytes, smooth muscle cells, and pericytes adhere (Varma et al. 2007). Within the $CD34^{+}$ cells, two subpopulations with different phenotypes have been identified (a $CD34^{dim}$ and $CD34^{bright}$ subpopulation) (Varma et al. 2007). In addition, no markers of macrophages, lymphocytes, or granulocytes could be detected in primary cultured ASCs by flow cytometry (Brzoska et al. 2005). Whereas the presence of endothelial cells is not detectable in primary cultured ASCs (Brzoska et al. 2005), it should be mentioned that endothelial cells in culture are extremely susceptible to culture conditions, such as supplements and particularly shear

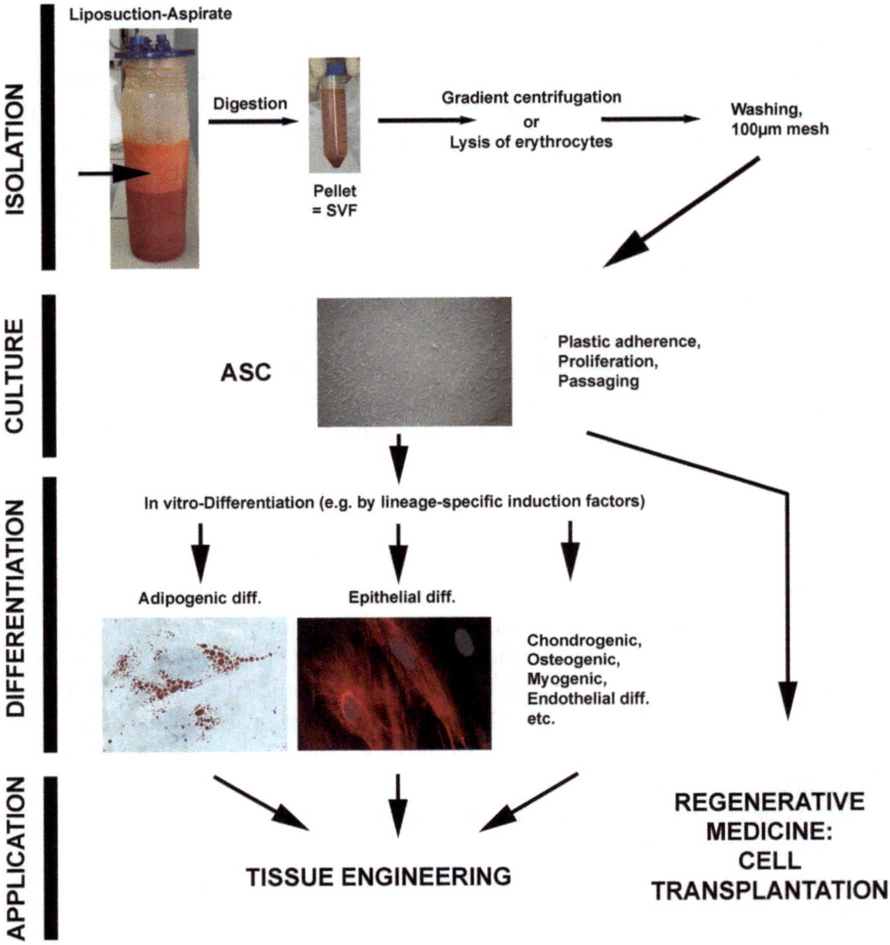

Fig. 4.1 Scheme of isolation, culture, differentiation and possible applications of ASCs. To isolate the stomal vascular fraction (SVF), the raw liposuction aspirate is washed extensively, and digested with 0.075 % collagenase. The digested fat is centrifuged to eliminate floating adipocytes, lipids and undigested adipose tissue. Contaminating red blood cells are eliminated either by lysis in NH_4Cl or by gradient centrifugation. After washing and centrifugation, the pellet is filtered through a 100 µm nylon mesh and resuspended in cell culture medium. ASCs are proliferated and passaged in culture. Differentiation could be achieved in vitro by lineage-specific induction factors. Adipogenic differentiation is shown by *Oil-O-Red* staining, epithelial differentiation is shown by *immunofluorescence* staining (cytokeratin 18 and nuclear contrasting)

stress. Endothelial cells dedifferentiate or trigger apoptosis under static culture conditions (Kaiser et al. 1997). Therefore, it would be valuable to remove contaminating $CD31^+$ endothelial cells in the primary isolates by negative selection. The heterogeneity of cultured ASCs can also be reduced by a washing procedure early in the beginning of the cell culture (Griesche et al. 2010), indicating that several subsets require different time points to adhere to the cell culture plastic.

Cultured ASCs show an extensive proliferative ability in an uncommitted state while retaining their multilineage differentiation potential.

In later passages, ASC cultures are homogeneous and exhibit a fibroblastoid morphology. The composition of subpopulations, therefore, may change during expansion (Schellenberg et al. 2012). Cell culture selects for this homogeneous morphology, enriching for cells expressing a stromal immunophenotype (Mitchell et al. 2006). Different studies have characterized and compared the immunophenotype of cultured ASCs in early and later passages over the past few years, and found that the expression profile of ASCs changes during culture time (Astori et al. 2007; Locke et al. 2009; Mitchell et al. 2006; Schellenberg et al. 2012). ASCs in passage 2 or later uniformly express their characteristic markers (positive for CD10, CD13, CD29, CD 44, CD73, CD90, CD105, and CD166, and negative for CD11b, CD14, CD31, and HLA-DR) (Gronthos et al. 2001). Furthermore, some reports described CD34, CD54, CD107, or CD146 to be expressed on cultured ASCs, and others did not find the expression of these antigens. These differing results are due to differences in the isolation or culture method, or caused by the investigation of different passages of cultured ASCs: For example, although CD34 is reckoned as a hematopoietic stem cell associated marker, it is expressed by early passages of ASCs' subsets and subsequently lost in later passages (Astori et al. 2007; Mitchell et al. 2006; Yoshimura et al. 2006). Discrepancies in the results of studies from different laboratories may result from many different origins: donor differencies, methods and quality of isolation, and whole culture procedure (e.g. medium, coating) (Baer 2011; Wagner et al. 2006; Wagner and Ho 2007). For example, the media composition highly effects the expression of the stem cell-related transcription factors NANOG, Oct-4, Sox-2, and Rex-1 in ASCs (Baer et al. 2010). In summary, modifications in the isolation and/or culture conditions might select for the expansion of subpopulations and have a huge impact on the differentiation potential of the cells cultured, albeit the primary cells could be phenotypically identical (Ho et al. 2008). Therefore, standardization of the laboratory techniques (isolation and culture) is highly needed for a good reproducibility of results from different laboratories and studies.

The Differentiation Potential of ASCs

Approximately 10 years ago, several publications reported on the ability of MSCs in general to differentiate across primary germ layer lineages and boundaries (Jiang et al. 2002; Lange et al. 2011). Recently, the *in vitro* differentiation of ASCs into multiple cell types of all three primary germ layers has also been shown. Therefore, the term pluripotent stem cells would be more correct for ASCs (rather than multipotent). Whereas many studies support the pluripotent potential of ASCs (reviewed in Schaffler and Buchler 2007; Zuk 2010) it is not overally accepted in the scientific community, because ASC morphology is different to other pluripotent stem cells, and their ability to form teratoma has not been shown. Further investigations are needed, also to compare ASCs transcriptom and their epigenetic signature with pluripotent embyonic stem cells. Nevertheless, the work of Peroni and coworkers clearly showed the "stemness" transcriptional profile of ASCs as well as bone marrow MSC (BM-MSCs), and a close similarity to embryonic stem cells (Peroni et al. 2008). Both ASCs and BM-MSCs express molecules that are typical of embryonic cells, such as Oct-4, the undifferentiation transcription factor-1 and Nodal (Peroni et al. 2008).

It seems to be obvious, that ASCs are able to differentiate into cells of mesodermal origin, because ASCs themselves are of mesodermal origin. Different studies described ASCs' plasticity towards adipocytes, osteoblasts, chondrocytes, myocytes (cardiomyocytes, smooth muscle and skeletal muscle cells), pericytes, tenocytes, and vascular endothelial cells (Feng et al. 2010, 2006; Gimble and Guilak 2003a, b; Lee and Kemp 2006; Mizuno et al. 2002; Planat-Benard et al. 2004; Rodriguez et al. 2006; Strem et al. 2005; Uysal and Mizuno 2011; Zuk et al. 2002, 2001). However, ASCs have also been able to differentiate into cells of ecto- or endodermal origin. The *in vitro* differentiation into neurons, oligodendrocytes, Schwann cells, and keratocytes has been shown (Ashjian et al. 2003; Du et al. 2010; Kingham et al. 2009; Safford et al. 2002). In addition, differentiation into hepatocytes, keratinocytes, pancreatic islet cells, and other epithelial cells has been described (Baer et al. 2009, 2011, 2013; Banas et al. 2007; Brzoska et al. 2005; Ebrahimian et al. 2009; Griesche et al. 2012; Lange et al. 2006; Long et al. 2010; Seo et al. 2005; Timper et al. 2006; Zuk 2010). Nonetheless, it should be mentioned that some epithelial cells arise from mesodermal origin (e.g. some nephronal parts of the kidney or vascular endothelial cells).

In general, the induction of ASCs' differentiation *in vitro* is mainly achieved by culture in selective media with lineage-specific induction factors. For example, our studies have clearly verified that ASCs can enter the epithelial lineage when treated with retinoids (Brzoska et al. 2005), conditioned medium (CM) from injured epithelial cells or a mixture of growth factors (Baer et al. 2009, 2011; Banas et al. 2007; Timper et al. 2006). The transcriptional and molecular events triggering the mesodermal lineage-specific differentiation of stem cells are well known (Davis and Zur Nieden 2008; James et al. 2010; Karbiener et al. 2009; Li et al. 2008; Liu et al. 2007; Schaffler and Buchler 2007), but little is known about the molecular mechanisms during differentiation into ecto- and endodermal lineages. ASCs have also been shown to be angiogenesis and hematopoiesis supporting cells (Corre et al. 2006; Cousin et al. 2003; Miranville et al. 2004). These supporting characteristics of ASCs are mainly due to the secretion of anti-apoptotic, angiogenic and hematopoietic factors (cytokines and growth factors), such as macrophage colony-stimulating factor, granulocyte macrophage colony-stimulating factor, insulin-like growth factor, hepatocyte growth factor, vascular endothelial growth factor, hepatocyte growth factor, and transforming growth factor-β (Kilroy et al. 2007; Nakagami et al. 2005; Rehman et al. 2004).

Several tissue engineering and cell therapeutical approaches using ASCs with or without scaffolds have been carried out in animal experiments to verify ASCs' *in vitro* differentiation potential. Besides the obvious applications of ASCs to repair or regenerate cartilage, bone, muscle, or adipose tissue, the possibility of peripheral nerve regeneration, hepatic regeneration, insulin-producing islet cell regeneration, functional repair of myocardial infarction, and recovery of renal function has recently been shown in *in vivo* models (Aurich et al. 2009; Banas et al. 2009; di Summa et al. 2010; Kajiyama et al. 2010; Li et al. 2010; Santiago et al. 2009; Schenke-Layland et al. 2009).

Endothelial Differentiation Potential of ASCs *in vitro* and *in vivo*

As described in the last chapter, the vascular endothelium developmentally originates from the embryonic mesoderm, and, therefore, shares the embryonal origin with ASCs. In the adult, differentiated vascular endothelial cells are recognized as specialized simple squamous epithelial cells.

Beginning with the first verifications of ASCs' differentiation capacity into endothelial cells in 2004 (Miranville et al. 2004; Planat-Benard et al. 2004), several in vitro and in vivo studies confirmed this differentiation potential (Table 4.1). In vitro, the induction is mostly carried out by the addition of growth factors and/or other lineage-specific factors to the culture medium, maybe in combination with culture plastic coating or shear force. The analysis of endothelial differentiation *in vitro* is verified by the ability of the cells to form vessel-like structures during the differentiation process and the expression of endothelial-specific markers. Von Willebrand factor (vWF), angiotensin converting enzyme (ACE), and vascular endothelial growth factor receptor-2 (VEGFR-2, Flk-1, KDR) are the earliest markers in the endothelial differentation program (Kajiyama et al. 2010; Schenke-Layland et al. 2009). Uptake of acetylated low density lipoprotein (acLDL) has also been observed early in endothelial differentiation. Other molecules expressed during onset of endothelial differentiation are VEGFR–1 (Flt-1), vascular endothelial cadherin (VE-cadherin), P-selectin, CD31, and CD34 (Risau 1995). VE-cadherin, VEGFR-1, and -2 are most specific for endothelial cells, whereas ACE, vWF, CD31, P-selectin, and CD34 are also present in other cell types (Risau 1995). Furthermore, it should be mentioned that a low-level expression of endothelial cell–associated markers (CD31, VE-cadherin, VEGFR-2, vWF) has been described on undifferentiated ASCs (Mitchell et al. 2006), whereas other studies did not confirm the expression of these markers.

In vivo studies revealed that systemically administered ASCs localize to the site of injury and contribute to revascularization of injured tissue. Endothelial differentiation of transplanted ASCs (either GFP-transgene or otherwise pre-labeled) was verified by fluorescence analysis of the labeled cells in combination with the analysis of an endothelial-specific marker (Bai et al. 2010; Moon et al. 2006; Nie et al. 2011; Ning et al. 2009; Uysal et al. 2009; Valina et al. 2007). Others used anti-human-specific antibodies to stain transplanted human ASCs incorporated within the mouse vasculature (Miranville et al. 2004; Planat-Benard et al. 2004). Furthermore, microangiography or a laser Doppler blood flow meter were used as indirect

Table 4.1 Summary of selected in vitro and in vivo studies describing endothelial differentiation of ASCs

Authors	Model	Cells	In vitro Stimulation/In vivo model	Results/Verification
Planat-Benard et al. (2004)	In vitro/In vivo	Murine and Human ASCs	Semisolid medium/Hindlimb ischemia, Mouse	CD31 ↑, vWF ↑, formation of tube-like structures Incorporation into vessels, neovascularization
Miranville et al. (2004)	In vitro/n vivo	Human ASCs[a]	Medium with VEGF and IGF/Hindlimb ischemia, Mouse	CD31 ↑, vWF ↑/Increased blood flow and capillary density, incorporation in the vasculature
Cao et al. (2005)	In vitro/In vivo	Human ASCs	Medium with VEGF and bFGF/Hindlimb ischemia, Mouse	CD31 ↑, CD34 ↑, eNOS ↑, VE-cadherin ↑, acLDL uptake/Neoangiogenesis
Martinez-Estrada et al. (2005)	In vitro	Human ASCs[b]	Specific Medium	vWF ↑, VE-cadherin ↑, lectin receptor ↑, acLDL uptake, incorporation in endothelial cell tubular networks.
Moon et al. (2006)	In vitro/In vivo	Human ASCs[c]	Substrate and medium/Hindlimb ischemia, Mouse	vWF ↑, formation of vessel-like structures/Increased vascular density, increased perfusion index
Lin et al. (2007)	In vitro	Murine ASCs	Gas-plasma treated bioabsorbable scaffolds and differentiation medium	VEGF release ↑, cyclooxygenase-2 ↑, matrix metalloproteinase-2 ↑
Wosnitza et al. (2007)	In vitro	Human ASCs[d]	Substrate and medium with VEGF and IGF	CD34 ↑, VEGF receptor (KDR) ↑, acLDL uptake
Valina et al. (2007)	In vivo	Porcine ASCs	Myocardial infarction, Porcine	vWF ↑ of transplanted labeled ASCs, incorporation into newly formed vessels,increased vascular density
Uysal et al. (2009)	In vivo	Murine ASCs	Skin flap ischemia, Mouse	vWF ↑ of transplanted GFP+ -ASCs, increased vascular density, increased number of vessels, increased levels of VEGF, TGFb, and FGF

Table 4.1 (continued)

Authors	Model	Cells	In vitro Stimulation/In vivo model	Results/Verification
Ning et al. (2009)	In vitro/In vivo	Rat ASCs	Medium with bFGF/Injection in corpora cavernosa, Rat	CD31 ↑, vWF ↑, eNOS ↑, acLDL uptake, formation of tube-like structures/RECA-1 expression of transplanted labeled ASCs
Fischer et al. (2009)	In vitro	Human ASCs Canine ASCs	Substrate, medium and shear stress	acLDL uptake, CD31 ↑, vWF ↑
Nie et al. (2010)	In vivo	Rat ASCs	Wound healing, Rat	CD31 ↑ of transplanted GFP+ -ASCs, Neovascularization
Bai et al. (2010)	In vivo	Human ASCs	Myocardial infarction, Mouse	vWF ↑ of transplanted labeled ASCs, increased vascular density
Marino et al. (2011)	In vitro	Human ASCs	Substrate and medium with bFGF	acLDL uptake, CD31 ↑, vWF ↑, eNOS ↑
Zhang et al. (2011)	In vitro	Human ASCs[e]	VEGF, shear stress	CD31 ↑, vWF ↑, eNOS ↑, VE-cadherin ↑, acLDL- and lectin-uptake

↑ De novo expression (or upregulation); ↓ Expression downregulated

acLDL Acetylated low-density lipoprotein; *bFGF* Basic fibroblast growth factor; *eNOS* Endothelial nitric oxide synthase; *IGF* Insulin-like growth factor; *RECA* Rat endothelial cell antigen; *VEGF* Vascular endothelial growth factor; *vWF* Van Willebrand factor

[a] Miranville et al. used two magnetic bead isolated subpopulations of ASCs ($CD34^+$ /$CD31^-$ and $CD34^+$ /$CD31^+$)

[b] Martinez-Estrada et al selected $Flk\text{-}1^+$ ASCs by culture conditions and medium

[c] Moon et al. used the magnetic bead isolated $CD34^+$ -subpopulation of ASCs

[d] Wosnitza et al. used two magnetic bead isolated subpopulations of ASCs ($CD31^+$ and $CD31^-$)

[e] Zhang et al. depleted the $CD31^+$ and $CD45^+$ cells by magnetic beads

techniques to validate the effect of transplanted cells (Cao et al. 2005; Miranville et al. 2004; Moon et al. 2006; Planat-Benard et al. 2004; Uysal et al. 2009). Miranville and co-workers demonstrated that intravenous injection of ASCs in mouse ischemic hind limb is followed by an increase in blood flow and capillary density and the incorporation of transplanted cells in the leg vasculature (Miranville et al. 2004). Planat-Benard and co-workers reported similar findings nearly at the same time (Planat-Benard et al. 2004). They demonstrated that human ASCs differentiate into endothelial cells *in vitro*, incorporate into vessels *in vivo* and promote postischemic neovascularization in mice (Planat-Benard et al. 2004). Fang and co-workers showed multiorgan engraftment of transplanted human ASCs and that the cells contribute to neoangiogenesis in a mouse model. They detected the cells in mouse vessels by immunofluorescence stainings with human-specific antibodies against vWF and the Y chromosome (Fang et al. 2010), and discriminated them from host-specific vWF^{+} endothelial cells (with a anti-mouse antibody). Uysal and co-workers observed in an ischemia-reperfusion model that approximately 21 % of the whole endothelial cells originated from donor-ASCs on postoperative day 7 (Uysal et al. 2009).

In addition, it has been described that CD13^{+}/CD34^{+} ASCs cultured in semisolid medium (Matrigel) spontaneously expressed CD31 and vWF, and formed vascular structures in vitro (Planat-Benard et al. 2004). It has also been demonstrated that a subpopulation of ASCs (CD34^{+}/CD31^{-} cells) is able to differentiate into endothelial cells if cultured in endothelial growth medium supplemented with insulin-like growth factor (IGF) and vascular endothelial growth factor (VEGF) (Miranville et al. 2004). In this study, endothelial differentiation was proven by upregulation of vWF and de novo expression of CD31. Cao and co-workers isolated a CD34^{-}/CD31^{-} subset from uncultured ASCs and showed that this subset expresses endothelial markers if cultured in the presence of VEGF (Cao et al. 2005). In vivo, the authors also demonstrated that this subset is able to contribute to neoangiogenesis in a hind limb ischemia model.

Ning and co-workers demonstrated that basic fibroblast growth factor (bFGF) promoted differentiation of ASCs into endothelial cells, and confirmed the differentiation by the verification of CD31, vWF, and endothelial nitric oxide synthase (eNOS) expression and the ability of forming vessel-like structures in Matrigel (Ning et al. 2009). Many other in vitro studies affirmed ASC's endothelial differentiation potential. Nevertheless, what is missing under routine endothelial cell culture methods is the shear force of flowing fluid. The influence of shear force on fully differentiated endothelial cells has been shown in many studies. Shear force has also been shown to promote endothelial differentiation of embryonic stem and endothelial progenitor cells (Wang et al. 2005; Yamamoto et al. 2005, 2003). The lack of shear force increases the rate of apoptosis, but this can be overcome by perfusion culture producing fluid flow over the cells (Freyberg et al. 2001a, b; Graf et al. 2003; Kaiser et al. 1997), maintaining characteristic endothelial differentiation markers. Furthermore, the expression of the Fas receptor is regulated by flow conditions in differentiated endothelial cells (Freyberg et al. 2001b). Fully differentiated endothelial cells cultivated under regular shear stress do not express the Fas receptor whereas cultivation under static conditions leads to its expression. Thus, endothelial cells cultivated under flow conditions are completely devoid of apoptosis, and apoptosis is induced by turning off the flow. Two recent studies analysed the effect of shear force of endothelial induced ASCs *in vitro* (Fischer et al. 2009; Zhang et al. 2011). The investigators used an orbital shaker to induce shear force rather than real perfusion culture in which laminar flow is generated and can be calculated precisely. However, it has been shown that shear force generated by orbital shaking upregulated CD31 expression and enhanced acetylated LDL uptake in ASCs pre-differentiated with endothelial growth medium (Fischer et al. 2009). Another study showed that shear force further stimulates endothelial differentiation of ASCs (Zhang et al. 2011). They showed that shear stress further upregulated endothelial cell-specific markers (CD31, vWF, endothelial nitric oxide synthase, and VE-cadherin). Shear stress also promoted an anti-thrombogenic phenotype as demonstrated by production of tissue-plasminogen activator and nitric oxide, and inhibition of plasminogen activator inhibitor-1, and augmented integrin $\alpha(5)\beta(1)$ expression (Zhang et al. 2011).

The molecular mechanisms underlying ASCs differentiation into the endothelial lineage are not fully understood. Only a few *in vitro* studies have been done to explore the signaling pathways, second messengers, and transcription factors related to ASCs endothelial differentiation. For example, the combination of VEGF and bFGF has been described to induce endothelial differentiation of ASCs (Cao et al. 2005). Inhibition of phosphatidylinositol-3-kinase by a specific inhibitor blocked the differentiation into endothelial cells *in vitro*, indicating that this pathway is essential for ASCs endothelial differentiation (Cao et al. 2005; Zhang et al. 2011). An inhibitor of the mitogen-activated protein kinase (MAPK) did not affect this differentiation (Cao et al. 2005; Zhang et al. 2011). Furthermore, Ning and co-workers showed that blockade of bFGF signaling with a specific inhibitor abrogated ASC's endothelial differentiation (Valina et al. 2007).

Final Remarks

ASCs are probably one of the most powerful adult stem cells. Apart from all the proven basic scientific evidence from *in vitro* and *in vivo* studies, which have been accumulated during the last few years, the exact mechanisms underlying the beneficial effects of transplanted ASCs remain unclear. Although the integration of ASCs into vessels and their *in vivo* differentiation into endothelial cells has been shown, they may mostly act through paracrine pathways via secreted cytoprotective cytokines and growth factors like VEGF, hepatocyte growth factor, or other angiogenic molecules. Maybe both mechanisms in concert are responsible for the ameliorating effects in neoangiogenesis or neovascularization of ischemia and other injuries. Recently (in May 2012), six clinical trials using culture-isolated ASCs to promote neoangiogenesis or induce neovascularization in ischemic diseases have been initiated (http://www.clinicaltrials.gov). The primary intent of these studies is safety, feasibility and effectiveness of the therapy rather than verification of the exact mechanisms of action.

In conclusion, the basic knowledge about ASCs needs to be enhanced. An important factor to improve the use of ASCs is the standardization in the isolation methods and culture protocols. Good manufacturing practices (GMP) qualified isolation and culture protocols are needed for the use of ASCs in clinical trials. Research progress has also been hampered by the limited knowledge of ASCs' subsets, due to the lack of unique markers for their isolation. Nevertheless, the entirety of recent *in vivo* studies and of the few published case reports and clinical trials have shown that ASCs are on the direct path to their clinical usage for the treatment of a multitude of diseases.

Acknowledgement Parts of this work were supported by the Adolf Messer Stiftung, Königstein, Germany.

References

Ashjian PH, Elbarbary AS, Edmonds B, DeUgarte D, Zhu M, Zuk PA, Lorenz HP, Benhaim P, Hedrick MH (2003) In vitro differentiation of human processed lipoaspirate cells into early neural progenitors. Plastic Reconstr Surg 111(6):1922–1931. doi:10.1097/01.PRS.0000055043.62589.05

Astori G, Vignati F, Bardelli S, Tubio M, Gola M, Albertini V, Bambi F, Scali G, Castelli D, Rasini V, Soldati G, Moccetti T (2007) In vitro and multicolor phenotypic characterization of cell subpopulations identified in fresh human adipose tissue stromal vascular fraction and in the derived mesenchymal stem cells. J Transl Mede 5:55. doi:10.1186/1479–5876-5–55

Aurich H, Sgodda M, Kaltwasser P, Vetter M, Weise A, Liehr T, Brulport M, Hengstler JG, Dollinger MM, Fleig WE, Christ B (2009) Hepatocyte differentiation of mesenchymal stem cells from human adipose tissue in vitro promotes hepatic integration in vivo. Gut 58(4):570–581. doi:10.1136/gut.2008.154880

Aust L, Devlin B, Foster SJ, Halvorsen YD, Hicok K, du Laney T, Sen A, Willingmyre GD, Gimble JM (2004) Yield of human adipose-derived adult stem cells from liposuction aspirates. Cytotherapy 6(1):7–14. doi:10.1080/14653240310004539

Baer PC (2011) Adipose-derived stem cells and their potential to differentiate into the epithelial lineage. Stem Cells Dev 20(10):1805–1816. doi:10.1089/scd.2011.0086

Baer PC, Bereiter-Hahn J, Missler C, Brzoska M, Schubert R, Gauer S, Geiger H (2009) Conditioned medium from renal tubular epithelial cells initiates differentiation of human mesenchymal stem cells. Cell Prolif 42(1):29–37. doi:10.1111/j.1365–2184.2008.00572.x

Baer PC, Griesche N, Luttmann W, Schubert R, Luttmann A, Geiger H (2010) Human adipose-derived mesenchymal stem cells in vitro: evaluation of an optimal expansion medium preserving stemness. Cytotherapy 12(1):96–106. doi:10.3109/14653240903377045

Baer PC, Brzoska M, Geiger H (2011) Epithelial differentiation of human adipose-derived stem cells. Methods Mol Biol 702:289–298. doi:10.1007/978–1-61737–960-4_21

Baer PC, Doring C, Hansmann ML, Schubert R, Geiger H (2013) New insights into epithelial differentiation of human adipose-derived stem cells. J Tissue Eng Regen Med 7(4):271–278. doi:10.1002/term.518

Bai X, Yan Y, Song YH, Seidensticker M, Rabinovich B, Metzele R, Bankson JA, Vykoukal D, Alt E (2010) Both cultured and freshly isolated adipose tissue-derived stem cells enhance cardiac

function after acute myocardial infarction. Eur Heart J 31(4):489–501. doi:10.1093/eurheartj/ehp568
Banas A, Teratani T, Yamamoto Y, Tokuhara M, Takeshita F, Quinn G, Okochi H, Ochiya T (2007) Adipose tissue-derived mesenchymal stem cells as a source of human hepatocytes. Hepatology 46(1):219–228. doi:10.1002/hep.21704
Banas A, Teratani T, Yamamoto Y, Tokuhara M, Takeshita F, Osaki M, Kato T, Okochi H, Ochiya T (2009) Rapid hepatic fate specification of adipose-derived stem cells and their therapeutic potential for liver failure. J Gastroenterol Hepatol 24(1):70–77. doi:10.1111/j.1440–1746. 2008.05496.x
Brzoska M, Geiger H, Gauer S, Baer P (2005) Epithelial differentiation of human adipose tissue-derived adult stem cells. Biochem Biophys Res Commun 330(1):142–150. doi:10.1016/j.bbrc.2005.02.141
Cai X, Lin Y, Hauschka PV, Grottkau BE (2011) Adipose stem cells originate from perivascular cells. Biol Cell 103(9):435–447. doi:10.1042/BC20110033
Cao Y, Sun Z, Liao L, Meng Y, Han Q, Zhao RC (2005) Human adipose tissue-derived stem cells differentiate into endothelial cells in vitro and improve postnatal neovascularization in vivo. Biochem Biophys Res Commun 332(2):370–379. doi:10.1016/j.bbrc.2005.04.135
Corre J, Barreau C, Cousin B, Chavoin JP, Caton D, Fournial G, Penicaud L, Casteilla L, Laharrague P (2006) Human subcutaneous adipose cells support complete differentiation but not self-renewal of hematopoietic progenitors. J Cell Physiol 208(2):282–288. doi:10.1002/jcp.20655
Corselli M, Chen CW, Crisan M, Lazzari L, Peault B (2010) Perivascular ancestors of adult multipotent stem cells. Arterioscler Thromb Vasc Biol 30(6):1104–1109. doi:10.1161/ATVBAHA.109.191643
Cousin B, Andre M, Arnaud E, Penicaud L, Casteilla L (2003) Reconstitution of lethally irradiated mice by cells isolated from adipose tissue. Biochem Biophys Res Commun 301(4):1016–1022
Crisan M, Yap S, Casteilla L, Chen CW, Corselli M, Park TS, Andriolo G, Sun B, Zheng B, Zhang L, Norotte C, Teng PN, Traas J, Schugar R, Deasy BM, Badylak S, Buhring HJ, Giacobino JP, Lazzari L, Huard J, Peault B (2008) A perivascular origin for mesenchymal stem cells in multiple human organs. Cell Stem Cell 3(3):301–313. doi:10.1016/j.stem.2008.07.003
Crisan M, Chen CW, Corselli M, Andriolo G, Lazzari L, Peault B (2009) Perivascular multipotent progenitor cells in human organs. Ann N Y Acad Sci 1176:118–123. doi:10.1111/j.1749–6632.2009.04967.x
da Silva Meirelles L, Chagastelles PC, Nardi NB (2006) Mesenchymal stem cells reside in virtually all post-natal organs and tissues. J Cell Sci 119(Pt 11):2204–2213. doi:10.1242/jcs.02932
Davis LA, Zur Nieden NI (2008) Mesodermal fate decisions of a stem cell: the Wnt switch. Cell Mol Life Sci 65(17):2658–2674. doi:10.1007/s00018–008-8042–1
De Ugarte DA, Morizono K, Elbarbary A, Alfonso Z, Zuk PA, Zhu M, Dragoo JL, Ashjian P, Thomas B, Benhaim P, Chen I, Fraser J, Hedrick MH (2003) Comparison of multi-lineage cells from human adipose tissue and bone marrow. Cells Tissues Organs 174(3):101–109. doi:71150
di Summa PG, Kingham PJ, Raffoul W, Wiberg M, Terenghi G, Kalbermatten DF (2010) Adipose-derived stem cells enhance peripheral nerve regeneration. J Plastic Reconstr Aesthet Surg 63(9):1544–1552. doi:10.1016/j.bjps.2009.09.012
Dominici M, Le Blanc K, Mueller I, Slaper-Cortenbach I, Marini F, Krause D, Deans R, Keating A, Prockop D, Horwitz E (2006) Minimal criteria for defining multipotent mesenchymal stromal cells. The International Society for Cellular Therapy position statement. Cytotherapy 8(4):315–317. doi:10.1080/14653240600855905
Dominici M, Paolucci P, Conte P, Horwitz EM (2009) Heterogeneity of multipotent mesenchymal stromal cells: from stromal cells to stem cells and vice versa. Transplantation 87(9 Suppl):S36–42. doi:10.1097/TP.0b013e3181a283ee
Du Y, Roh DS, Funderburgh ML, Mann MM, Marra KG, Rubin JP, Li X, Funderburgh JL (2010) Adipose-derived stem cells differentiate to keratocytes in vitro. Mol Vis 16:2680–2689
Ebrahimian TG, Pouzoulet F, Squiban C, Buard V, Andre M, Cousin B, Gourmelon P, Benderitter M, Casteilla L, Tamarat R (2009) Cell therapy based on adipose tissue-derived stromal

cells promotes physiological and pathological wound healing. Arterioscler Thromb Vasc Biol 29(4):503–510. doi:10.1161/ATVBAHA.108.178962

Fang B, Li Y, Song Y, Li N, Cao Y, Wei X, Lin Q, Zhao RC (2010) Human adipose tissue-derived adult stem cells can lead to multiorgan engraftment. Transplant Proc 42(5):1849–1856. doi:10.1016/j.transproceed.2010.01.058

Feng J, Mantesso A, Sharpe PT (2010) Perivascular cells as mesenchymal stem cells. Expert Opin Biol Ther 10(10):1441–1451. doi:10.1517/14712598.2010.517191

Fischer LJ, McIlhenny S, Tulenko T, Golesorkhi N, Zhang P, Larson R, Lombardi J, Shapiro I, DiMuzio PJ (2009) Endothelial differentiation of adipose-derived stem cells: effects of endothelial cell growth supplement and shear force. J Surg Res 152(1):157–166. doi:10.1016/j.jss.2008.06.029

Fraser JK, Schreiber R, Strem B, Zhu M, Alfonso Z, Wulur I, Hedrick MH (2006) Plasticity of human adipose stem cells toward endothelial cells and cardiomyocytes. Nat Clin Prac Cardiovasc Med 3(Suppl 1):S33–37. doi:10.1038/ncpcardio0444

Fraser JK, Zhu M, Wulur I, Alfonso Z (2008) Adipose-derived stem cells. Methods Mol Biol 449:59–67. doi:10.1007/978–1-60327–169-1_4

Freyberg MA, Kaiser D, Graf R, Buttenbender J, Friedl P (2001a) Proatherogenic flow conditions initiate endothelial apoptosis via thrombospondin-1 and the integrin-associated protein. Biochem Biophys Res Commun 286(1):141–149. doi:10.1006/bbrc.2001.5314

Freyberg MA, Kaiser D, Graf R, Friedl P (2001b) Vascular endothelial cells express a functional fas-receptor due to lack of hemodynamic forces. Apoptosis 6(5):339–343

Gimble J, Guilak F (2003a) Adipose-derived adult stem cells: isolation, characterization, and differentiation potential. Cytotherapy 5(5):362–369. doi:10.1080/14653240310003026

Gimble JM, Guilak F (2003b) Differentiation potential of adipose derived adult stem (ADAS) cells. Curr Top Dev Biol 58:137–160

Graf R, Apenberg S, Freyberg M, Friedl P (2003) A common mechanism for the mechanosensitive regulation of apoptosis in different cell types and for different mechanical stimuli. Apoptosis 8(5):531–538

Griesche N, Luttmann W, Luttmann A, Stammermann T, Geiger H, Baer PC (2010) A simple modification of the separation method reduces heterogeneity of adipose-derived stem cells. Cells Tissues Organs 192(2):106–115. doi:10.1159/000289586

Griesche N, Bereiter-Hahn J, Geiger H, Schubert R, Baer PC (2012) During epithelial differentiation of human adipose-derived stromal/stem cells, expression of zonula occludens protein-1 is induced by a combination of retinoic acid, activin-A and bone morphogenetic protein-7. Cytotherapy 14(1):61–69. doi:10.3109/14653249.2011.610502

Grimes BR, Steiner CM, Merfeld-Clauss S, Traktuev DO, Smith D, Reese A, Breman AM, Thurston VC, Vance GH, Johnstone BH, Slee RB, March KL (2009) Interphase FISH demonstrates that human adipose stromal cells maintain a high level of genomic stability in long-term culture. Stem Cells Dev 18(5):717–724. doi:10.1089/scd.2008.0255

Gronthos S, Franklin DM, Leddy HA, Robey PG, Storms RW, Gimble JM (2001) Surface protein characterization of human adipose tissue-derived stromal cells. J Cell Physiol 189(1):54–63. doi:10.1002/jcp.1138

Ho AD, Wagner W, Franke W (2008) Heterogeneity of mesenchymal stromal cell preparations. Cytotherapy 10(4):320–330. doi:10.1080/14653240802217011

Hoogduijn MJ, Roemeling-van Rhijn M, Korevaar SS, Engela AU, Weimar W, Baan CC (2011) Immunological aspects of allogeneic and autologous mesenchymal stem cell therapies. Hum Gene Ther 22(12):1587–1591. doi:10.1089/hum.2011.039

James AW, Leucht P, Levi B, Carre AL, Xu Y, Helms JA, Longaker MT (2010) Sonic Hedgehog influences the balance of osteogenesis and adipogenesis in mouse adipose-derived stromal cells. Tissue Eng Part A 16(8):2605–2616. doi:10.1089/ten.TEA.2010.0048

Jiang Y, Jahagirdar BN, Reinhardt RL, Schwartz RE, Keene CD, Ortiz-Gonzalez XR, Reyes M, Lenvik T, Lund T, Blackstad M, Du J, Aldrich S, Lisberg A, Low WC, Largaespada DA, Verfaillie CM (2002) Pluripotency of mesenchymal stem cells derived from adult marrow. Nature 418(6893):41–49. doi:10.1038/nature00870

Kaiser D, Freyberg MA, Friedl P (1997) Lack of hemodynamic forces triggers apoptosis in vascular endothelial cells. Biochem Biophys Res Commun 231(3):586–590. doi:10.1006/bbrc.1997.6146

Kajiyama H, Hamazaki TS, Tokuhara M, Masui S, Okabayashi K, Ohnuma K, Yabe S, Yasuda K, Ishiura S, Okochi H, Asashima M (2010) Pdx1-transfected adipose tissue-derived stem cells differentiate into insulin-producing cells in vivo and reduce hyperglycemia in diabetic mice. Int J Dev Biol 54(4):699–705. doi:10.1387/ijdb.092953hk

Karbiener M, Fischer C, Nowitsch S, Opriessnig P, Papak C, Ailhaud G, Dani C, Amri EZ, Scheideler M (2009) microRNA miR-27b impairs human adipocyte differentiation and targets PPARgamma. Biochem Biophys Res Commun 390(2):247–251. doi:10.1016/j.bbrc.2009.09.098

Kassis I, Zangi L, Rivkin R, Levdansky L, Samuel S, Marx G, Gorodetsky R (2006) Isolation of mesenchymal stem cells from G-CSF-mobilized human peripheral blood using fibrin microbeads. Bone Marrow Transplant 37(10):967–976. doi:10.1038/sj.bmt.1705358

Kilroy GE, Foster SJ, Wu X, Ruiz J, Sherwood S, Heifetz A, Ludlow JW, Stricker DM, Potiny S, Green P, Halvorsen YD, Cheatham B, Storms RW, Gimble JM (2007) Cytokine profile of human adipose-derived stem cells: expression of angiogenic, hematopoietic, and pro-inflammatory factors. J Cell Physiol 212(3):702–709. doi:10.1002/jcp.21068

Kingham PJ, Mantovani C, Terenghi G (2009) Notch independent signalling mediates Schwann cell-like differentiation of adipose derived stem cells. Neurosci Lett 467(2):164–168. doi:10.1016/j.neulet.2009.10.030

Lange C, Bruns H, Kluth D, Zander AR, Fiegel HC (2006) Hepatocytic differentiation of mesenchymal stem cells in cocultures with fetal liver cells. World J Gastroenterol 12(15):2394–2397

Lange C, Brunswig-Spickenheier B, Cappallo-Obermann H, Eggert K, Gehling UM, Rudolph C, Schlegelberger B, Cornils K, Zustin J, Spiess AN, Zander AR (2011) Radiation rescue: mesenchymal stromal cells protect from lethal irradiation. PloS one 6(1):e14486. doi:10.1371/journal.pone.0014486

Lee JH, Kemp DM (2006) Human adipose-derived stem cells display myogenic potential and perturbed function in hypoxic conditions. Biochem Biophys Res Commun 341(3):882–888. doi:10.1016/j.bbrc.2006.01.038

Li HX, Luo X, Liu RX, Yang YJ, Yang GS (2008) Roles of Wnt/beta-catenin signaling in adipogenic differentiation potential of adipose-derived mesenchymal stem cells. Mol Cell Endocrinol 291(1–2):116–124. doi:10.1016/j.mce.2008.05.005

Li K, Han Q, Yan X, Liao L, Zhao RC (2010) Not a process of simple vicariousness, the differentiation of human adipose-derived mesenchymal stem cells to renal tubular epithelial cells plays an important role in acute kidney injury repairing. Stem Cells Dev 19(8):1267–1275. doi:10.1089/scd.2009.0196

Lin G, Garcia M, Ning H, Banie L, Guo YL, Lue TF, Lin CS (2008) Defining stem and progenitor cells within adipose tissue. Stem Cells Dev 17(6):1053–1063. doi:10.1089/scd.2008.0117

Lin CS, Xin ZC, Deng CH, Ning H, Lin G, Lue TF (2010) Defining adipose tissue-derived stem cells in tissue and in culture. Histol Histopathol 25(6):807–815

Liu TM, Martina M, Hutmacher DW, Hui JH, Lee EH, Lim B (2007) Identification of common pathways mediating differentiation of bone marrow- and adipose tissue-derived human mesenchymal stem cells into three mesenchymal lineages. Stem Cells 25(3):750–760. doi:10.1634/stemcells.2006–0394

Locke M, Windsor J, Dunbar PR (2009) Human adipose-derived stem cells: isolation, characterization and applications in surgery. ANZ J Surg 79(4):235–244. doi:10.1111/j.1445–2197.2009.04852.x

Long JL, Zuk P, Berke GS, Chhetri DK (2010) Epithelial differentiation of adipose-derived stem cells for laryngeal tissue engineering. Laryngoscope 120(1):125–131. doi:10.1002/lary.20719

McIntosh K, Zvonic S, Garrett S, Mitchell JB, Floyd ZE, Hammill L, Kloster A, Di Halvorsen Y, Ting JP, Storms RW, Goh B, Kilroy G, Wu X, Gimble JM (2006) The immunogenicity of human adipose-derived cells: temporal changes in vitro. Stem Cells 24(5):1246–1253. doi:10.1634/stemcells.2005–0235

McIntosh KR (2011) Evaluation of cellular and humoral immune responses to allogeneic adipose-derived stem/stromal cells. Methods Mol Biol 702:133–150. doi:10.1007/978-1-61737-960-4_11

Miranville A, Heeschen C, Sengenes C, Curat CA, Busse R, Bouloumie A (2004) Improvement of postnatal neovascularization by human adipose tissue-derived stem cells. Circulation 110(3):349–355. doi:10.1161/01.CIR.0000135466.16823.D0

Mitchell JB, McIntosh K, Zvonic S, Garrett S, Floyd ZE, Kloster A, Di Halvorsen Y, Storms RW, Goh B, Kilroy G, Wu X, Gimble JM (2006) Immunophenotype of human adipose-derived cells: temporal changes in stromal-associated and stem cell-associated markers. Stem Cells 24(2):376–385. doi:10.1634/stemcells.2005–0234

Mizuno H, Zuk PA, Zhu M, Lorenz HP, Benhaim P, Hedrick MH (2002) Myogenic differentiation by human processed lipoaspirate cells. Plast Reconstr Surg 109(1):199–209; discussion 210–191

Moon MH, Kim SY, Kim YJ, Kim SJ, Lee JB, Bae YC, Sung SM, Jung JS (2006) Human adipose tissue-derived mesenchymal stem cells improve postnatal neovascularization in a mouse model of hindlimb ischemia. Cellular Physiol Biochem 17(5–6):279–290. doi:10.1159/000094140

Nakagami H, Maeda K, Morishita R, Iguchi S, Nishikawa T, Takami Y, Kikuchi Y, Saito Y, Tamai K, Ogihara T, Kaneda Y (2005) Novel autologous cell therapy in ischemic limb disease through growth factor secretion by cultured adipose tissue-derived stromal cells. Arterioscler Thromb Vasc Biol 25(12):2542–2547. doi:10.1161/01.ATV.0000190701.92007.6d

Nie C, Yang D, Xu J, Si Z, Jin X, Zhang J (2011) Locally administered adipose-derived stem cells accelerate wound healing through differentiation and vasculogenesis. Cell Transplant 20(2):205–216. doi:10.3727/096368910X520065

Ning H, Liu G, Lin G, Yang R, Lue TF, Lin CS (2009) Fibroblast growth factor 2 promotes endothelial differentiation of adipose tissue-derived stem cells. J Sex Med 6(4):967–979. doi:10.1111/j.1743–6109.2008.01172.x

Oedayrajsingh-Varma MJ, van Ham SM, Knippenberg M, Helder MN, Klein-Nulend J, Schouten TE, Ritt MJ, van Milligen FJ (2006) Adipose tissue-derived mesenchymal stem cell yield and growth characteristics are affected by the tissue-harvesting procedure. Cytotherapy 8(2):166–177. doi:10.1080/14653240600621125

Peroni D, Scambi I, Pasini A, Lisi V, Bifari F, Krampera M, Rigotti G, Sbarbati A, Galie M (2008) Stem molecular signature of adipose-derived stromal cells. Exp Cell Res 314(3):603–615. doi:10.1016/j.yexcr.2007.10.007

Planat-Benard V, Silvestre JS, Cousin B, Andre M, Nibbelink M, Tamarat R, Clergue M, Manneville C, Saillan-Barreau C, Duriez M, Tedgui A, Levy B, Penicaud L, Casteilla L (2004) Plasticity of human adipose lineage cells toward endothelial cells: physiological and therapeutic perspectives. Circulation 109(5):656–663. doi:10.1161/01.CIR.0000114522.38265.61

Rehman J, Traktuev D, Li J, Merfeld-Clauss S, Temm-Grove CJ, Bovenkerk JE, Pell CL, Johnstone BH, Considine RV, March KL (2004) Secretion of angiogenic and antiapoptotic factors by human adipose stromal cells. Circulation 109(10):1292–1298. doi:10.1161/01.CIR.0000121425.42966.F1

Risau W (1995) Differentiation of endothelium. FASEB J 9(10):926–933

Rodriguez LV, Alfonso Z, Zhang R, Leung J, Wu B, Ignarro LJ (2006) Clonogenic multipotent stem cells in human adipose tissue differentiate into functional smooth muscle cells. Proc Nat Acad Sci U S A 103(32):12167–12172. doi:10.1073/pnas.0604850103

Safford KM, Hicok KC, Safford SD, Halvorsen YD, Wilkison WO, Gimble JM, Rice HE (2002) Neurogenic differentiation of murine and human adipose-derived stromal cells. Biochem Biophys Res Commun 294 (2):371–379. doi:10.1016/S0006–291X(02)00469–2

Santiago LY, Clavijo-Alvarez J, Brayfield C, Rubin JP, Marra KG (2009) Delivery of adipose-derived precursor cells for peripheral nerve repair. Cell Transplant 18(2):145–158

Schaffler A, Buchler C (2007) Concise review: adipose tissue-derived stromal cells–basic and clinical implications for novel cell-based therapies. Stem Cells 25(4):818–827. doi:10.1634/stemcells.2006–0589

Schellenberg A, Stiehl T, Horn P, Joussen S, Pallua N, Ho AD, Wagner W (2012) Population dynamics of mesenchymal stromal cells during culture expansion. Cytotherapy 14(4):401–411. doi:10.3109/14653249.2011.640669

Schenke-Layland K, Strem BM, Jordan MC, Deemedio MT, Hedrick MH, Roos KP, Fraser JK, Maclellan WR (2009) Adipose tissue-derived cells improve cardiac function following myocardial infarction. J Surg Res 153(2):217–223. doi:10.1016/j.jss.2008.03.019

Seo MJ, Suh SY, Bae YC, Jung JS (2005) Differentiation of human adipose stromal cells into hepatic lineage in vitro and in vivo. Biochem Biophys Res Commun 328(1):258–264. doi:10.1016/j.bbrc.2004.12.158

Strem BM, Hicok KC, Zhu M, Wulur I, Alfonso Z, Schreiber RE, Fraser JK, Hedrick MH (2005) Multipotential differentiation of adipose tissue-derived stem cells. Keio J Med 54(3):132–141

Timper K, Seboek D, Eberhardt M, Linscheid P, Christ-Crain M, Keller U, Muller B, Zulewski H (2006) Human adipose tissue-derived mesenchymal stem cells differentiate into insulin, somatostatin, and glucagon expressing cells. Biochem Biophys Res Commun 341(4):1135–1140. doi:10.1016/j.bbrc.2006.01.072

Traktuev DO, Prater DN, Merfeld-Clauss S, Sanjeevaiah AR, Saadatzadeh MR, Murphy M, Johnstone BH, Ingram DA, March KL (2009) Robust functional vascular network formation in vivo by cooperation of adipose progenitor and endothelial cells. Circ Res 104(12):1410–1420. doi:10.1161/CIRCRESAHA.108.190926

Uysal AC, Mizuno H (2011) Differentiation of adipose-derived stem cells for tendon repair. Methods Mol Biol 702:443–451. doi:10.1007/978-1-61737–960-4_32

Uysal AC, Mizuno H, Tobita M, Ogawa R, Hyakusoku H (2009) The effect of adipose-derived stem cells on ischemia-reperfusion injury: immunohistochemical and ultrastructural evaluation. Plast Reconstr Surg 124(3):804–815. doi:10.1097/PRS.0b013e3181b17bb4

Valina C, Pinkernell K, Song YH, Bai X, Sadat S, Campeau RJ, Le Jemtel TH, Alt E (2007) Intracoronary administration of autologous adipose tissue-derived stem cells improves left ventricular function, perfusion, and remodelling after acute myocardial infarction. Eur Heart J 28(21):2667–2677. doi:10.1093/eurheartj/ehm426

Varma MJ, Breuls RG, Schouten TE, Jurgens WJ, Bontkes HJ, Schuurhuis GJ, van Ham SM, van Milligen FJ (2007) Phenotypical and functional characterization of freshly isolated adipose tissue-derived stem cells. Stem Cells Dev 16(1):91–104. doi:10.1089/scd.2006.0026

Wagner W, Feldmann RE Jr, Seckinger A, Maurer MH, Wein F, Blake J, Krause U, Kalenka A, Burgers HF, Saffrich R, Wuchter P, Kuschinsky W, Ho AD (2006) The heterogeneity of human mesenchymal stem cell preparations–evidence from simultaneous analysis of proteomes and transcriptomes. Exp Hematol 34(4):536–548. doi:10.1016/j.exphem.2006.01.002

Wagner W, Ho AD (2007) Mesenchymal stem cell preparations–comparing apples and oranges. Stem Cell Rev 3(4):239–248. doi:10.1007/s12015–007-9001–1

Wang H, Riha GM, Yan S, Li M, Chai H, Yang H, Yao Q, Chen C (2005) Shear stress induces endothelial differentiation from a murine embryonic mesenchymal progenitor cell line. Arterioscler Thromb Vasc Biol 25(9):1817–1823. doi:10.1161/01.ATV.0000175840.90510.a8

Yamamoto K, Takahashi T, Asahara T, Ohura N, Sokabe T, Kamiya A, Ando J (2003) Proliferation, differentiation, and tube formation by endothelial progenitor cells in response to shear stress. J Appl Physiol 95(5):2081–2088. doi:10.1152/japplphysiol.00232.2003

Yamamoto K, Sokabe T, Watabe T, Miyazono K, Yamashita JK, Obi S, Ohura N, Matsushita A, Kamiya A, Ando J (2005) Fluid shear stress induces differentiation of Flk-1-positive embryonic stem cells into vascular endothelial cells in vitro. Am J Physiol Heart Circ Physiol 288(4):H1915–1924. doi:10.1152/ajpheart.00956.2004

Yoshimura K, Shigeura T, Matsumoto D, Sato T, Takaki Y, Aiba-Kojima E, Sato K, Inoue K, Nagase T, Koshima I, Gonda K (2006) Characterization of freshly isolated and cultured cells derived from the fatty and fluid portions of liposuction aspirates. J Cell Physiol 208(1):64–76. doi:10.1002/jcp.20636

Zannettino AC, Paton S, Arthur A, Khor F, Itescu S, Gimble JM, Gronthos S (2008) Multipotential human adipose-derived stromal stem cells exhibit a perivascular phenotype in vitro and in vivo. J Cell Physiol 214(2):413–421. doi:10.1002/jcp.21210

Zhang P, Moudgill N, Hager E, Tarola N, Dimatteo C, McIlhenny S, Tulenko T, DiMuzio PJ (2011) Endothelial differentiation of adipose-derived stem cells from elderly patients with cardiovascular disease. Stem Cells Dev 20(6):977–988. doi:10.1089/scd.2010.0152

Zhu Y, Liu T, Song K, Fan X, Ma X, Cui Z (2008) Adipose-derived stem cell: a better stem cell than BMSC. Cell Biochem Funct 26(6):664–675. doi:10.1002/cbf.1488

Zimmerlin L, Donnenberg VS, Pfeifer ME, Meyer EM, Peault B, Rubin JP, Donnenberg AD (2010) Stromal vascular progenitors in adult human adipose tissue. Cytometry A 77(1):22–30. doi:10.1002/cyto.a.20813

Zou Z, Zhang Y, Hao L, Wang F, Liu D, Su Y, Sun H (2010) More insight into mesenchymal stem cells and their effects inside the body. Expert Opin Biol Ther 10(2):215–230. doi:10.1517/14712590903456011

Zuk PA (2010) The adipose-derived stem cell: looking back and looking ahead. Mol Bio Cell 21(11):1783–1787. doi:10.1091/mbc.E09–07-0589

Zuk PA, Zhu M, Mizuno H, Huang J, Futrell JW, Katz AJ, Benhaim P, Lorenz HP, Hedrick MH (2001) Multilineage cells from human adipose tissue: implications for cell-based therapies. Tissue Eng 7(2):211–228. doi:10.1089/107632701300062859

Zuk PA, Zhu M, Ashjian P, De Ugarte DA, Huang JI, Mizuno H, Alfonso ZC, Fraser JK, Benhaim P, Hedrick MH (2002) Human adipose tissue is a source of multipotent stem cells. Mol Bio Cell 13(12):4279–4295. doi:10.1091/mbc.E02–02-0105

Chapter 5
Cardiac Cell Replacement Therapy With Pluripotent Stem Cell Derived Cardiomyocytes

Benjamin Krausgrill, Marcel Halbach and Jürgen Hescheler

Abstract This chapter tries to give an overview on cardiac cell replacement therapy with pluripotent stem cell derived cardiomyocytes. After a general introduction addressing concept, potential cell types and potential routes of administration, it highlights electrophysiology as well as engraftment and persistence of transplanted cardiomyocytes summarizing what has been achieved and what still needs further investigations.

Cardiac Cell Replacement Therapy: Concept, Potential Cell Types and Potential Routes of Administration

The Concept of Cardiac Cell Replacement Therapy

According to the current data of the World Health Organization, most of the deaths in the world are caused by ischaemic heart disease with a total number of 7.25 million or 12.8% of all deaths in 2008 (World Health Organization: http://www.who.int). Ischaemic heart disease can be subdivided into the acute phase of an acute myocardial infarction and the chronic phase of chronic coronary artery disease and chronic ischaemic cardiomyopathy. In an acute myocardial infarction, a pre-existing atheriosclerotic plaque inside the wall of a coronary artery ruptures and therefore activates local blood coagulation with thrombocyte adhesion and aggregation, which finally leads to a thrombotic occlusion of the coronary artery. The resulting stop of blood flow causes a lack of oxygen and nutrition as well as a pondage of metabolic waste in the downstream myocardium where cardiomyocytes then time-dependently die-off. Since thereby lost contractile myocardium is replaced by stiff scar tissue, an impaired pump function of the heart remains which then is called ischaemic cardiomyopathy (Fig. 5.1).

J. Hescheler (✉) · B. Krausgrill · M. Halbach
Institute of Neurophysiology, University Hospital of Cologne,
Robert-Koch-Str. 39, 50931, Cologne, Germany
e-mail: j.hescheler@uni-koeln.de

J. Hescheler, E. Hofer (eds.), *Adult and Pluripotent Stem Cells*,
DOI 10.1007/978-94-017-8657-7_5, © Springer Science+Business Media Dordrecht 2014

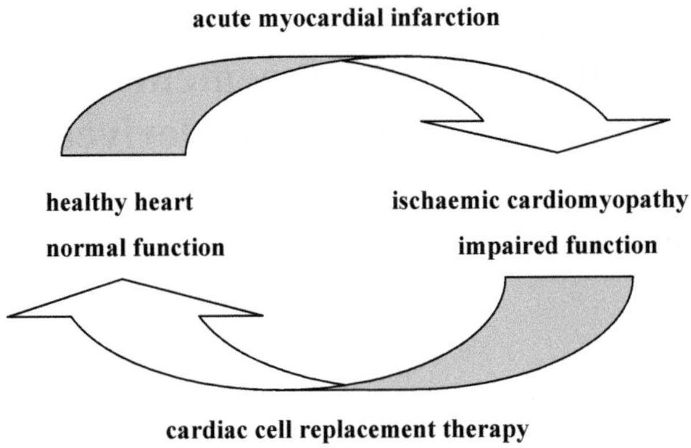

Fig. 5.1 The concept of cardiac cell replacement therapy

Over the last decades, new pharmacological and interventional therapies lead to a better and much faster reperfusion of the infarcted myocardium, which decreased mortality of the acute event. However and somehow in consequence, there is a high and still growing burden of patients with the chronic phase of ischaemic cardiomyopathy. In this situation, there is no or only very limited regenerative potential within the damaged heart whose function will therefore be impaired forever. Thus, the concept of cardiac cell replacement therapy was developed with the idea to replace the scar area of lost myocardium with transplanted new cells in order to regain a normal heart function. In the end, these transplanted new cells should be contractile and form a functional syncytium with the host myocardium to really normalize physiological heart function.

Potential Cell Types for Cardiac Cell Replacement Therapy

Although not contractile, cells from bone-marrow or blood are widely used for cardiac cell replacement therapy both in the experimental and in the clinical setting. One of the reasons is a ground-breaking paper of Orlic et al. published in Nature in 2001 (Orlic et al. 2001a). After complete replacement of the original wild-type bone-marrow with GFP-positive bone-marrow cells, acute myocardial infarction was induced in mice and thereafter GFP-positive cells were found in the infarcted hearts. This homing of bone-marrow cells into the damaged myocardium was thought to have regenerative potential which might become effective in healing myocardial infarction when homing was increased. Consequently, Orlic et al. mobilized bone-marrow cells into the blood of infarcted mice and found significant regeneration with decreased infarct size, improved heart function and decreased mortality (Orlic et al. 2001b). Although transdifferentiation of bone-marrow cells into cardiomyocytes was shown in histology and immunohistochemistry in this and other studies, other groups could not find evidence for such a transdifferentiation of bone-marrow cells into cardiomyocytes (Balsam et al. 2004; Nygren et al. 2004) which is still highly controversial. Nevertheless, cells from bone-marrow or blood

Table 5.1 Potential cell types for cardiac cell replacement therapy

Cell types	Non-contractile		Contractile	
	Bone-marrow or blood cells	Pluripotent stem cells (e.g. ES or iPS)	Skeletal myoblasts	Cardiomyocytes
Autologous transplantation possible?	Yes	Yes (for iPS only)	Yes	Yes
(Trans-) Differentiation to cardiomyocytes possible?	No	Yes	No	Not needed
Isolation/	Yes/	No/	Yes/	No/
Culture/	Yes/	Yes/	Yes/	No/
Upscaling easy?	To some extent	Yes	To some extent	No

were widely used in clinical trials and some of which showed beneficial effects e.g. (Assmus et al. 2006; Piepoli et al. 2010; Schachinger et al. 2006; Yao et al. 2009; van Ramshorst et al. 2009), while others did not (Herbots et al. 2009; Janssens et al. 2006; Lunde et al. 2006; Meyer et al. 2006; Traverse et al. 2011). While these different findings were discussed to be caused by slight differences in cell isolation and culture (and therefore in cell type) used, meta-analyses clearly demonstrated a marginal but significant improvement in heart function after treatment with cells from bone-marrow or blood (Jeevanantham et al. 2012). However, a real replacement of scar tissue with transplanted new contractile cells has not been achieved with these cells whose transplantation therefore cannot be called cardiac cell replacement therapy.

In other studies on cardiac cell replacement therapy, skeletal myoblasts were used. These cells are good candidates because they have the proven potential to differentiate into contractile cells. However, electromechanical coupling of these cells with host myocardium was either absent or malfunctioning or caused dangerous arrhythmias of the whole heart (for review Durrani et al. 2010). Thus, cardiomyocytes are needed to achieve real cardiac cell replacement therapy (Table 5.1).

Seeking cells with proven potential to differentiate into cardiomyocytes, many investigators focussed on pluripotent stem cells. Embryonic stem cells (ES) are truly pluripotent and can give rise to all kinds of cells during differentiation, which typically leads to teratoma formation after transplantation of undifferentiated embryonic stem cells. Although depending on cell dose and immunological matching of donor and recipient (Krausgrill et al. 2008, p. 370), such teratoma formation was also shown to happen in infarcted hearts after cardiac cell replacement therapy with undifferentiated embryonic stem cells which therefore could not be helpful. The same holds true for the transplantation of undifferentiated induced pluripotent stem cells (iPS).

In consequence, early or adult cardiomyocytes or their direct precursors without teratogenic risk are the best candidates for cardiac cell replacement therapy. Such cells can be derived from pluripotent stem cells (embryonic stem cells or induced pluripotent stem cells), but they must be highly purified in order not to co-transplant also teratogenic undifferentiated pluripotent stem cells.

Potential Routes of Administration for Cardiac Cell Replacement Therapy

In general, there are several potential routes of administration for cardiac cell replacement therapy, which include intravascular infusion (in peripheral veins, in coronary arteries or in cardiac veins), intramyocardial injection (transepicardially in open-chest surgery or transendocardially in cardiac catheterization) or deposition in the pericardial space. Each options offer both advantages and disadvantages:

Infusions into peripheral veins are very easy to perform and routinely done for a variety of therapies. However, cells infused into a peripheral vein will be taken with the blood stream into the cavities of the right heart and then into the capillaries in the lung, where most of them will most likely be stuck without ever reaching the target infarct zone in the ventricular wall. Notwithstanding, studies with bone-marrow mobilization or peripheral infusion could show some homing of the cells into the heart. But the effectiveness of these strategies is very unlikely to ever enable numbers of new cells in the heart that could be called cardiac cell replacement therapy. The intracoronary infusion was first described by Strauer et al. (2001) and is widely used in clinical trials, although an arterial access is needed and a cardiac catheterization has to be performed. In order to give the cells more time to leave the vessel and enter the tissue of the ventricular wall, studies usually apply the stop-flow technique with an over-the-wire balloon catheter, which means that a balloon occludes the coronary artery for some minutes while the injection is done in the stop-flow area behind the balloon. After some minutes of reperfusion, the next cycle of stop-flow and cell infusion is performed. Interestingly, when exactly transferring the clinical trials' situation to a pig model of reperfused acute myocardial infarction, the retention of intracoronarily infused cells was similar with or without balloon occlusion of the coronary artery during cell infusion and it was dramatically lower when compared with direct transepicardial intramyocardial injection (Tossios et al. 2008). Since some washout of the cells was shown to happen although the balloon is still effectively occluding the coronary artery, a new strategy of occluding the cardiac veins during cell infusion in the stop-flow area in front of the balloon is investigated (Table 5.2).

In terms of efficient deposition, engraftment and retention of transplanted cells, intramyocardial injection remains the superior technique for cardiac cell replacement therapy. While the surgical transepicardial approach requires an open-chest surgery with respective efforts and risks, the interventional transendocardial approach needs an arterial access and a cardiac catheterization with some kind of mapping of the ventricular wall to localize target spots for cell injections.

An interesting new route of administration for cardiac cell replacement therapy is the deposition of cells into the pericardial space, which can be performed with an echo-guided subxiphoidal puncture of the pericardial sac and was shown to lead to efficient cell deposition and engraftment into the ventricular wall (Ladage et al. 2012).

Table 5.2 Potential routes of administration for cardiac cell replacement therapy

Route of administration	Intravascular		Intramyocardial		Pericardial
	Intravenous	Intracoronary	Transepicardial	Transendocardial	
Typical procedure needed	Regular venous puncture	Cardiac catheterization	Open-chest surgery	Cardiac catheterization	Subxiphoidal puncture
Invasivity	Very low	High	Very high	High	Low
Engraftment efficiency	Poor	Poor	High	High	High

Cardiac Cell Replacement Therapy with Pluripotent Stem Cell Derived Cardiomyocytes: Promising Electrophysiology but Challenging Engraftment

First of all, physiological properties are a prerequisite for an efficient, valid and safe use of pluripotent stem cell derived cardiomyocytes. Besides specific cardiac molecular markers and spontaneous beating activity, cardiomyocytes derived from embryonic stem cells (ES) and induced pluripotent stem cells (iPS) have electrophysiological properties similar to those of native cardiomyocytes. Using current differentiation protocols, pluripotent stem cell-derived cardiomyocytes resemble cardiomyocytes of an embryonic stage, with proven capability of subtype differentiation, intercellular coupling and intact vegetative regulation.

An important feature of native cardiomyocytes is subtype differentiation, mainly into atrial, ventricular and pacemaker cells. Several publications demonstrated that cardiac differentiation of murine embryonic stem cells leads to different phenotypes with currents and action potential morphologies specific for the three main cardiac subtypes (Doss et al. 2007; Kolossov et al. 2005; Maltsev et al. 1993, 1994).

Embryonic stem cell derived cardiomyocytes with properties of Purkinje cells, another cardiac subtype belonging to the conduction system, have been described, too (Wobus et al. 1997). The potential of subtype differentiation could also be shown for cardiomyocytes derived from human embryonic stem cells (He et al. 2003; Mummery et al. 2003), from murine induced pluripotent stem cells (Kuzmenkin et al. 2009) and from human induced pluripotent stem cells (Zhang et al. 2009). The percentage of atrial, ventricular and pacemaker cells varies. It has been described that different cells in one embryoid body have quite homogenous electrophysiological properties, while properties of different embryoid bodies vary, giving rise to the cardiac subtypes (Zhang et al. 2009). Besides action potentials typical for the cardiac subtypes, unspecific and atypical action potential morphologies have been found, too (Doss et al. 2007; Maltsev et al. 1993).

Heart rate is controlled by the vegetative nervous system *in vivo*, and an intact response to vegetative regulation is another prerequisite for pluripotent stem cell derived cardiomyocytes in cardiac cell replacement therapy. In murine and human embryonic stem cell derived cardiomyocytes, beating frequency is regulated by adrenergic and muscarinic signalling in a physiological way (Banach et al. 2003;

Maltsev et al. 1999; Reppel et al. 2004). All elements of the ß-adrenergic signalling cascade become active during development of embryonic stem cell derived cardiomyocytes (Maltsev et al. 1999). In induced pluripotent stem cell derived cardiomyocytes, a chronotropic effect of adrenergic and muscarinic stimulation was shown, too (Pfannkuche et al. 2009; Zhang et al. 2009).

During native murine development, action potential morphologies and the underlying currents change substantially. While action potentials of early stage ventricular cardiomyocytes have little phase 1 repolarisation but a pronounced plateau phase and action potential durations to 50% repolarisation of approximately 100–200 ms (Fijnvandraat et al. 2003), the plateau is diminished during further fetal and postnatal development. Adult murine cardiomyocytes have a strong phase 1 repolarisation leading to a short action potential duration to 50% repolarisation of approximately 5 ms (Halbach et al. 2006). Moreover, diastolic potential becomes more stable and adult ventricular cardiomyocytes do not have the spontaneous beating activity of fetal cardiomyocytes.

In murine embryonic stem cell derived cardiomyocytes, action potential duration decreases during long-term culture, which may be interpreted as a sign of maturation (Banach et al. 2003; Fijnvandraat et al. 2003). Action potential duration of murine induced pluripotent stem cell derived cardiomyocytes also decreases during time in culture (Pfannkuche et al. 2009). However up to now, cardiomyocytes derived from embryonic stem cells or from induced pluripotent stem cells could never be differentiated to an adult phenotype. Even after long-term culture, overall action potential morphology, action potential duration, upstroke velocity and maximum diastolic potential are indicative of very immature cardiomyocytes (Fijnvandraat et al. 2003). Late stage murine embryonic stem cell derived cardiomyocytes most closely match day 12.5 embryonic cardiomyocytes of the atrium or outflow tract (Fijnvandraat et al. 2003). In contrast to native cardiomyocytes, the beating frequency of embryonic stem cell derived cardiomyocytes increases during time in culture (Banach et al. 2003; Fijnvandraat et al. 2003), affirming the limited—or even pointing to an atypical—developmental capacity, at least *in vitro*.

The immature developmental stage of pluripotent stem cell derived cardiomyocytes may be explained with the need for special differentiation factors or niches, which cannot (or not yet) be reproduced *in vitro*. However, it cannot be excluded that the developmental potential of cardiomyocytes derived from embryonic stem cells or from induced pluripotent stem cells may be generally compromised and that they cannot differentiate terminally—maybe even not under conditions after transplantation.

Not only an intact electrophysiology at the single cell level, but also a physiological conduction at tissue level is crucial for efficiency and safety of pluripotent stem cell derived cardiomyocytes in cardiac cell replacement therapy. The expression of connexins has been demonstrated in murine and human embryonic stem cell derived cardiomyocytes (Kolossov et al. 2006; Mummery et al. 2003) and murine induced pluripotent stem cell derived cardiomyocytes (Pfannkuche et al. 2009) by immunostainings. Moreover, spread of electrical excitation has been measured inside clusters of cardiomyocytes derived from murine embryonic stem cells and from murine induced pluripotent stem cells (Pfannkuche et al. 2009) as well as between embryoid bodies and native cardiomyocytes *in vitro* by microelectrode arrays (Kehat et al. 2004; Xue et al. 2005). The velocity of excitation spread has been calculated to be approximately 0.25 m/s (Pfannkuche et al. 2009) which is similar to native fetal tissue (Pillekamp et al. 2006). In adult tissue, conduction velocity is approximately 1 m/s (Halbach et al. 2006). Another evidence for electrical coupling of embryonic stem cell derived cardiomyocytes and native myocardium was shown in *in vivo* transplantation studies in pig and guinea pig hearts, which could be paced by transplanted human embryonic stem cell derived cardiomyocytes (Kehat et al. 2004; Xue et al. 2005).

After transplantation into adult mouse hearts, native fetal cardiomyocytes which do not differentiate to an adult phenotype *in vitro* have been shown to maturate and integrate as indicated by changes in action potential morphology and physiological conduction of excitations from host myocardium (Halbach et al. 2007) and these processes depended on each other and on time after transplantation (Halbach et al. 2012). Similarly, integration into host myocardium could also be demonstrated for transplanted cardiomyocytes derived from induced pluripotent stem cells, but its quality was lower with more frequent conduction blocks (Halbach et al. 2013). Future studies will have to show if the *in vivo* maturation seen in transplanted fetal cardiomyocytes is also possible in transplanted ES- or iPS-derived cardiomyocytes.

Electrophysiological properties of cardiomyocytes derived from embryonic stem cells and from induced pluripotent stem cells have been compared in several studies. Zhang et al. found that both, human embryonic and induced pluripotent stem cells were able to generate cardiomyocytes of the three major subtypes (Zhang et al. 2009). In both cases, the ventricular phenotype was prevailing and beating rate, action potential duration, amplitude, upstroke velocity and maximum diastolic potential were similar (Zhang et al. 2009).

Kuzmenkin et al. and Pfannkuche et al. described a prevailing ventricular phenotype in cardiomyocytes derived from murine induced pluripotent stem cells in contrast to an atrial phenotype in cardiomyocytes derived from murine embryonic stem cells (Kuzmenkin et al. 2009; Pfannkuche et al. 2009). Accordingly, action potential duration was shorter in cardiomyocytes derived from embryonic stem cells than in those derived from induced pluripotent stem cells. Sodium and L-type calcium currents were similar in both cell types, potassium currents were not analyzed in detail. However, a direct comparison of action potential properties of cardiomyocytes derived from murine embryonic stem cells and from murine induced pluripotent stem cells is hampered by pronounced developmental alterations in action potential properties and a delay of cardiac development in cardiomyocytes derived from induced pluripotent stem cells as compared to cardiomyocytes derived from embryonic stem cells (Pfannkuche et al. 2009). Since differences between cardiomyocytes derived from different embryonic stem cell lines have been reported (Halbach et al. 2010, p. 322), the differences between cardiomyocytes derived from embryonic stem cells and from induced pluripotent stem cells are difficult to interpret and further analyses of cardiomyocytes derived from different lines of different pluripotent stem cells are needed for a reliable comparison of their electrophysiological properties.

Because of the dose-dependent teratogenic risk of undifferentiated pluripotent stem cells (Krausgrill et al. 2008, p. 370), they have to be differentiated

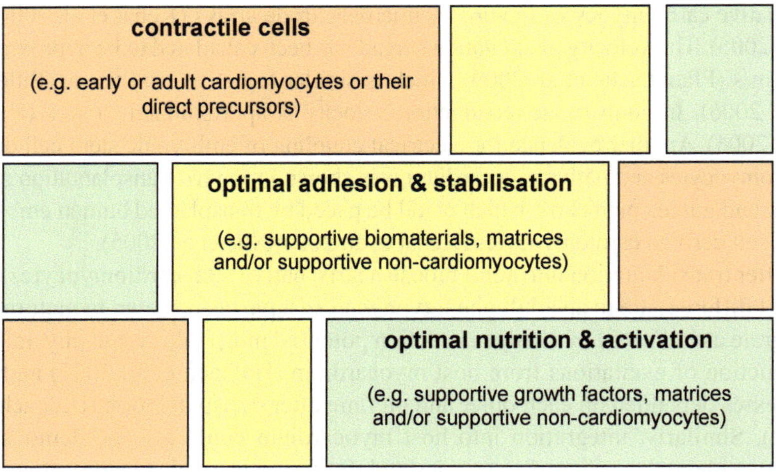

Fig. 5.2 The multifactorial approach of in-vivo tissue-engineering for cardiac cell replacement therapy

into early or adult cardiomyocytes or their direct precursors and they need to be highly purified prior to their transplantation in order to diminish the teratogenic risk of contaminating undifferentiated pluripotent stem cells. However, the purity of transplanted cardiomyocyte suspensions was shown to account for engraftment and persistence of the transplanted cells, which was increased when using an optimal—but not 100 %—purity of cardiomyocytes (Krausgrill et al. 2011a, p. 814). This optimal purity could be achieved by admixture of different non-cardiomyocytes, of which the usage of mesenchymal bone-marrow cells seemed to have additional benefits (Krausgrill et al. 2011b). Accordingly, highly purified cardiomyocytes derived from embryonic stem cells have a very limited persistence after transplantation in both healthy and injured hearts (Krausgrill et al. 2009, p. 498). Interestingly, the massive cell loss occurs already during or shortly after intramyocardial injection into mice hearts and could be dose-dependently delayed but not prevented by co-injection of Matrigel™ as an adhesive extracellular matrix (Krausgrill et al. 2010, p. 399). Highly purified cardiomyocytes derived from induced pluripotent stem cells show similarly poor persistence and although admixture of non-cardiomyocytes increases engraftment and persistence of these pluripotent stem cell derived cardiomyocytes, too, numbers are still by far too low in terms of a cardiac cell replacement therapy. Most likely, simple injections of only the cells needed will not be sufficient since they need optimal adhesion and stabilisation (e.g. through co-transplantation of supportive biomaterials, matrices and/or supportive non-cardiomyocytes), as well as optimal nutrition and activation (e.g. through co-transplantation of supportive growth factors and/or supportive non-cardiomyocytes) in a multifactorial approach of *in vivo* tissue engineering (Fig. 5.2).

Conclusion

Cardiac cell replacement therapy with pluripotent stem cell derived cardiomyocytes is a very promising approach to treat heart failure, since it aims at the root of the disease and tries to replace lost contractile cells with new ones in order to regain a normal cardiac function. To achieve this, real cardiomyocytes need to be transplanted into the failing heart in a sufficient number and with a sufficient engraftment efficiency and they need to be or become mature and functioning cardiomyocytes with sufficient electromechanical coupling into the host myocardium. Although to date cardiomyocytes derived from induced pluripotent stem cells are the most promising cell type to transplant, they still implicate some obstacles which need to be overcome like e.g. teratogenic risk due to insufficient purification or differentiation, insufficient engraftment and persistence, or insufficient maturation and coupling. The multifactorial approach of *in vivo* tissue-engineering for cardiac cell replacement therapy might be able to solve these problems and facilitate the desirable therapeutic benefits.

Source of funding This work was supported by an EC grant (FP7 HEALTH-F2-2009-222995 INELPY) to JH.

References

Assmus B, Honold J, Schachinger V, Britten MB, Fischer-Rasokat U, Lehmann R, Teupe C, Pistorius K, Martin H, Abolmaali ND, Tonn T, Dimmeler S, Zeiher AM (2006) Transcoronary transplantation of progenitor cells after myocardial infarction. New Eng J Med 355(12):1222–1232. doi:10.1056/NEJMoa051779

Balsam LB, Wagers AJ, Christensen JL, Kofidis T, Weissman IL, Robbins RC (2004) Haematopoietic stem cells adopt mature haematopoietic fates in ischaemic myocardium. Nature 428(6983):668–673. doi:10.1038/nature02460

Banach K, Halbach MD, Hu P, Hescheler J, Egert U (2003) Development of electrical activity in cardiac myocyte aggregates derived from mouse embryonic stem cells. Am J Physiol Heart Circ Physiol 284(6):H2114–2123. doi:10.1152/ajpheart.01106.2001

Doss MX, Winkler J, Chen S, Hippler-Altenburg R, Sotiriadou I, Halbach M, Pfannkuche K, Liang H, Schulz H, Hummel O, Hubner N, Rottscheidt R, Hescheler J, Sachinidis A (2007) Global transcriptome analysis of murine embryonic stem cell-derived cardiomyocytes. Genome Biol 8(4):R56. doi:10.1186/gb-2007-8-4-r56

Durrani S, Konoplyannikov M, Ashraf M, Haider KH (2010) Skeletal myoblasts for cardiac repair. Regen Med 5(6):919–932. doi:10.2217/rme.10.65

Fijnvandraat AC, van Ginneken AC, de Boer PA, Ruijter JM, Christoffels VM, Moorman AF, Lekanne Deprez RH (2003) Cardiomyocytes derived from embryonic stem cells resemble cardiomyocytes of the embryonic heart tube. Cardiovas Res 58(2):399–409

Halbach M, Pillekamp F, Brockmeier K, Hescheler J, Muller-Ehmsen J, Reppel M (2006) Ventricular slices of adult mouse hearts–a new multicellular in vitro model for electrophysiological studies. Cell Physiol Biochem 18(1–3):1–8. doi:10.1159/000095132

Halbach M, Pfannkuche K, Pillekamp F, Ziomka A, Hannes T, Reppel M, Hescheler J, Muller-Ehmsen J (2007) Electrophysiological maturation and integration of murine fetal cardiomyocytes after transplantation. Circ Res 101(5):484–492. doi:10.1161/CIRCRESAHA.107.153643

Halbach M, Krausgrill B, Hannes T, Wiedey M, Peinkofer G, Baumgartner S, Sahito RG, Pfannkuche K, Pillekamp F, Reppel M, Müller-Ehmsen J, Hescheler J (2012) Time-course of the electrophysiological maturation and integration of transplanted cardiomyocytes. J Mol Cell Cardiol 53(3):401–408. doi: 10.1016/j.yjmcc.2012.06.007. Epub 2012 Jun 21

Halbach M, Peinkofer G, Baumgartner S, Maass M, Wiedey M, Neef K, Krausgrill B, Ladage D, Fatima A, Saric T, Hescheler J, Müller-Ehmsen J (2013) Electrophysiological integration and action potential properties of transplanted cardiomyocytes derived from induced pluripotent stem cells. Cardiovasc Res 100(3):432–440. doi: 10.1093/cvr/cvt213. Epub 2013 Sep 16

Halbach M, Hannes T, Khalil M et al (2010) Functional diversity of cardiomyocytes derived from different embryonic stem cell lines. Eur Heart J 31(Supplement: 1):322

He JQ, Ma Y, Lee Y, Thomson JA, Kamp TJ (2003) Human embryonic stem cells develop into multiple types of cardiac myocytes: action potential characterization. Circ Res 93(1):32–39. doi:10.1161/01.RES.0000080317.92718.99

Herbots L, D'Hooge J, Eroglu E, Thijs D, Ganame J, Claus P, Dubois C, Theunissen K, Bogaert J, Dens J, Kalantzi M, Dymarkowski S, Bijnens B, Belmans A, Boogaerts M, Sutherland G, Van de Werf F, Rademakers F, Janssens S (2009) Improved regional function after autologous bone marrow-derived stem cell transfer in patients with acute myocardial infarction: a randomized, double-blind strain rate imaging study. Eur Heart J 30(6):662–670. doi:10.1093/eurheartj/ehn532

Janssens S, Dubois C, Bogaert J, Theunissen K, Deroose C, Desmet W, Kalantzi M, Herbots L, Sinnaeve P, Dens J, Maertens J, Rademakers F, Dymarkowski S, Gheysens O, Van Cleemput J, Bormans G, Nuyts J, Belmans A, Mortelmans L, Boogaerts M, Van de Werf F (2006) Autologous bone marrow-derived stem-cell transfer in patients with ST-segment elevation myocardial infarction: double-blind, randomised controlled trial. Lancet 367(9505):113–121. doi:10.1016/S0140-6736(05)67861-0

Jeevanantham V, Butler M, Saad A, Abdel-Latif A, Zuba-Surma EK, Dawn B (2012) Adult bone marrow cell therapy improves survival and induces long-term improvement in cardiac parameters: a systematic review and meta-analysis. Circulation 126(5):551–568. doi:10.1161/CIRCULATIONAHA.111.086074

Kehat I, Khimovich L, Caspi O, Gepstein A, Shofti R, Arbel G, Huber I, Satin J, Itskovitz-Eldor J, Gepstein L (2004) Electromechanical integration of cardiomyocytes derived from human embryonic stem cells. Nat Biotechnol 22(10):1282–1289. doi:10.1038/nbt1014

Kolossov E, Lu Z, Drobinskaya I, Gassanov N, Duan Y, Sauer H, Manzke O, Bloch W, Bohlen H, Hescheler J, Fleischmann BK (2005) Identification and characterization of embryonic stem cell-derived pacemaker and atrial cardiomyocytes. FASEB J 19(6):577–579. doi:10.1096/fj.03-1451fje

Kolossov E, Bostani T, Roell W, Breitbach M, Pillekamp F, Nygren JM, Sasse P, Rubenchik O, Fries JW, Wenzel D, Geisen C, Xia Y, Lu Z, Duan Y, Kettenhofen R, Jovinge S, Bloch W, Bohlen H, Welz A, Hescheler J, Jacobsen SE, Fleischmann BK (2006) Engraftment of engineered ES cell-derived cardiomyocytes but not BM cells restores contractile function to the infarcted myocardium. J Exp Med 203(10):2315–2327. doi:10.1084/jem.20061469

Krausgrill B, Schenk K, Raths M et al (2008) Survival, proliferation and solid tumor formation after injection of undifferentiated embryonic stem cells in infarcted mouse hearts depends on cell number and immunosuppression. Eur Heart J 29 (Supplement: 1):370

Krausgrill B, Halbach M, Soemantri SP et al (2009) Mid-term cell loss of mouse embryonic stem cell derived cardiomyocytes after intramyocardial injection into cryo-injured hearts is similarly high, but occurs earlier than in sham-operated mouse hearts. Eur Heart J 30 (Supplement: 1):498

Krausgrill B, Halbach M, Soemantri SP et al (2010) Loss of purified cardiomyocytes derived from embryonic stem cells after transplantation into syngeneic mouse hearts occurs early and can be delayed but not prevented by coinjection of Matrigel™. Eur Heart J 31 (Supplement: 1):399

Krausgrill B, Oeckenpoehler S, Maass, M et al (2011a) Influence of non-cardiomyocytes on persistence and survival of rat neonatal cardiomyocytes after injection into infarcted syngeneic hearts. Eur Heart J 32 (Supplement: 1):814

Krausgrill B, Oeckenpoehler S, Maass, M et al (2011b) Persistence and survival of rat neonatal cardiomyocytes after injection into infarcted syngeneic hearts depends on their purity and is improved by co-injection of mesenchymal bone-marrow cells. Circulation 124 (21, Supplement S); Meeting Abstract: A15620

Kuzmenkin A, Liang H, Xu G, Pfannkuche K, Eichhorn H, Fatima A, Luo H, Saric T, Wernig M, Jaenisch R, Hescheler J (2009) Functional characterization of cardiomyocytes derived from murine induced pluripotent stem cells in vitro. FASEB J 23(12):4168–4180. doi:10.1096/fj.08-128546

Ladage D, Ishikawa K, Tilemann L, Muller-Ehmsen J, Kawase Y (2012) Percutaneous methods of vector delivery in preclinical models. Gene Ther 19(6):637–641. doi:10.1038/gt.2012.14

Lunde K, Solheim S, Aakhus S, Arnesen H, Abdelnoor M, Egeland T, Endresen K, Ilebekk A, Mangschau A, Fjeld JG, Smith HJ, Taraldsrud E, Grogaard HK, Bjornerheim R, Brekke M, Muller C, Hopp E, Ragnarsson A, Brinchmann JE, Forfang K (2006) Intracoronary injection of mononuclear bone marrow cells in acute myocardial infarction. New Eng J Med 355(12):1199–1209. doi:10.1056/NEJMoa055706

Maltsev VA, Rohwedel J, Hescheler J, Wobus AM (1993) Embryonic stem cells differentiate in vitro into cardiomyocytes representing sinusnodal, atrial and ventricular cell types. Mech Dev 44(1):41–50

Maltsev VA, Wobus AM, Rohwedel J, Bader M, Hescheler J (1994) Cardiomyocytes differentiated in vitro from embryonic stem cells developmentally express cardiac-specific genes and ionic currents. Circ Res 75(2):233–244

Maltsev VA, Ji GJ, Wobus AM, Fleischmann BK, Hescheler J (1999) Establishment of beta-adrenergic modulation of L-type Ca2+ current in the early stages of cardiomyocyte development. Circ Res 84(2):136–145

Meyer GP, Wollert KC, Lotz J, Steffens J, Lippolt P, Fichtner S, Hecker H, Schaefer A, Arseniev L, Hertenstein B, Ganser A, Drexler H (2006) Intracoronary bone marrow cell transfer after myocardial infarction: eighteen months' follow-up data from the randomized, controlled BOOST (BOne marrOw transfer to enhance ST-elevation infarct regeneration) trial. Circulation 113(10):1287–1294. doi:10.1161/CIRCULATIONAHA.105.575118

Mummery C, Ward-van Oostwaard D, Doevendans P, Spijker R, van den Brink S, Hassink R, van der Heyden M, Opthof T, Pera M, de la Riviere AB, Passier R, Tertoolen L (2003) Differentiation of human embryonic stem cells to cardiomyocytes: role of coculture with visceral endoderm-like cells. Circulation 107(21):2733–2740. doi:10.1161/01.CIR.0000068356.38592.68

Nygren JM, Jovinge S, Breitbach M, Sawen P, Roll W, Hescheler J, Taneera J, Fleischmann BK, Jacobsen SE (2004) Bone marrow-derived hematopoietic cells generate cardiomyocytes at a low frequency through cell fusion, but not transdifferentiation. Nat Med 10(5):494–501. doi:10.1038/nm1040

Orlic D, Kajstura J, Chimenti S, Jakoniuk I, Anderson SM, Li B, Pickel J, McKay R, Nadal-Ginard B, Bodine DM, Leri A, Anversa P (2001a) Bone marrow cells regenerate infarcted myocardium. Nature 410(6829):701–705. doi:10.1038/35070587

Orlic D, Kajstura J, Chimenti S, Limana F, Jakoniuk I, Quaini F, Nadal-Ginard B, Bodine DM, Leri A, Anversa P (2001b) Mobilized bone marrow cells repair the infarcted heart, improving function and survival. Proc Nat Acad Sci U S A 98(18):10344–10349. doi:10.1073/pnas.181177898

Pfannkuche K, Liang H, Hannes T, Xi J, Fatima A, Nguemo F, Matzkies M, Wernig M, Jaenisch R, Pillekamp F, Halbach M, Schunkert H, Saric T, Hescheler J, Reppel M (2009) Cardiac myocytes derived from murine reprogrammed fibroblasts: intact hormonal regulation, cardiac ion channel expression and development of contractility. Cell Physiol Biochem 24(1–2):73–86. doi:10.1159/000227815

Piepoli MF, Vallisa D, Arbasi M, Cavanna L, Cerri L, Mori M, Passerini F, Tommasi L, Rossi A, Capucci A (2010) Bone marrow cell transplantation improves cardiac, autonomic, and functional indexes in acute anterior myocardial infarction patients (Cardiac Study). Eur J Heart Fail 12(2):172–180. doi:10.1093/eurjhf/hfp183

Pillekamp F, Reppel M, Brockmeier K, Hescheler J (2006) Impulse propagation in late-stage embryonic and neonatal murine ventricular slices. J Electrocardiol 39(4):425 e421–424. doi:10.1016/j.jelectrocard.2006.02.008

Reppel M, Boettinger C, Hescheler J (2004) Beta-adrenergic and muscarinic modulation of human embryonic stem cell-derived cardiomyocytes. Cell Physiol Biochem 14(4–6):187–196. doi:10.1159/000080326

Schachinger V, Erbs S, Elsasser A, Haberbosch W, Hambrecht R, Holschermann H, Yu J, Corti R, Mathey DG, Hamm CW, Suselbeck T, Assmus B, Tonn T, Dimmeler S, Zeiher AM (2006) Intracoronary bone marrow-derived progenitor cells in acute myocardial infarction. New Eng J Med 355(12):1210–1221. doi:10.1056/NEJMoa060186

Strauer BE, Brehm M, Zeus T, Gattermann N, Hernandez A, Sorg RV, Kogler G, Wernet P (2001) [Intracoronary, human autologous stem cell transplantation for myocardial regeneration following myocardial infarction]. Dtsch Med Wochenschr 126(34–35):932–938. doi:10.1055/s-2001-16579-2

Tossios P, Krausgrill B, Schmidt M, Fischer T, Halbach M, Fries JW, Fahnenstich S, Frommolt P, Heppelmann I, Schmidt A, Schomacker K, Fischer JH, Bloch W, Mehlhorn U, Schwinger RH, Muller-Ehmsen J (2008) Role of balloon occlusion for mononuclear bone marrow cell deposition after intracoronary injection in pigs with reperfused myocardial infarction. Eur Heart J 29(15):1911–1921. doi:10.1093/eurheartj/ehn218

Traverse JH, Henry TD, Ellis SG, Pepine CJ, Willerson JT, Zhao DX, Forder JR, Byrne BJ, Hatzopoulos AK, Penn MS, Perin EC, Baran KW, Chambers J, Lambert C, Raveendran G, Simon DI, Vaughan DE, Simpson LM, Gee AP, Taylor DA, Cogle CR, Thomas JD, Silva GV, Jorgenson BC, Olson RE, Bowman S, Francescon J, Geither C, Handberg E, Smith DX, Baraniuk S, Piller LB, Loghin C, Aguilar D, Richman S, Zierold C, Bettencourt J, Sayre SL, Vojvodic RW, Skarlatos SI, Gordon DJ, Ebert RF, Kwak M, Moye LA, Simari RD (2011) Effect of intracoronary delivery of autologous bone marrow mononuclear cells 2 to 3 weeks following acute myocardial infarction on left ventricular function: the LateTIME randomized trial. JAMA 306(19):2110–2119. doi:10.1001/jama.2011.1670

van Ramshorst J, Bax JJ, Beeres SL, Dibbets-Schneider P, Roes SD, Stokkel MP, de Roos A, Fibbe WE, Zwaginga JJ, Boersma E, Schalij MJ, Atsma DE (2009) Intramyocardial bone marrow cell injection for chronic myocardial ischemia: a randomized controlled trial. JAMA 301(19):1997–2004. doi:10.1001/jama.2009.685

Wobus AM, Kaomei G, Shan J, Wellner MC, Rohwedel J, Ji G, Fleischmann B, Katus HA, Hescheler J, Franz WM (1997) Retinoic acid accelerates embryonic stem cell-derived cardiac differentiation and enhances development of ventricular cardiomyocytes. J Mol Cell Cardio 29(6):1525–1539

Xue T, Cho HC, Akar FG, Tsang SY, Jones SP, Marban E, Tomaselli GF, Li RA (2005) Functional integration of electrically active cardiac derivatives from genetically engineered human embryonic stem cells with quiescent recipient ventricular cardiomyocytes: insights into the development of cell-based pacemakers. Circulation 111(1):11–20. doi:10.1161/01.CIR.0000151313.18547.A2

Yao K, Huang R, Sun A, Qian J, Liu X, Ge L, Zhang Y, Zhang S, Niu Y, Wang Q, Zou Y, Ge J (2009) Repeated autologous bone marrow mononuclear cell therapy in patients with large myocardial infarction. Eur J Heart Fail 11(7):691–698. doi:10.1093/eurjhf/hfp062

Zhang J, Wilson GF, Soerens AG, Koonce CH, Yu J, Palecek SP, Thomson JA, Kamp TJ (2009) Functional cardiomyocytes derived from human induced pluripotent stem cells. Circ Res 104(4):e30–41. doi:10.1161/CIRCRESAHA.108.192237

Chapter 6
Biomaterials for Cardiac Tissue Engineering and Regeneration

Emil Ruvinov and Smadar Cohen

Abstract Tissue regeneration after myocardial infarction (MI) represents a major challenge in cardiovascular therapy, as current clinical approaches are limited in their ability to regenerate damaged myocardium. This chapter presents an overview of two emerging strategies based on the use of biomaterials, as stand-alone therapy or in combination with regeneration signals and cells, for regenerating the infarcted heart. One strategy is cardiac tissue engineering, which creates cardiac patches from functional cells seeded in a biomaterial scaffold, bio-inspired to provide the appropriate interface for cellular interactions. Implementation of perfusion bioreactors and pre-vascularization strategies can nowadays produce thicker cardiac patches that are better integrated into the host, thus advancing the realization of this strategy in the clinics. The second strategy, injection of biomaterials, has shown great promise as a stand-alone therapy. The intracoronary delivery of alginate solution that undergoes gelation only at the infarct in the presence of elevated calcium ions, was shown to increase scar thickness and LV dimensions in acute MI models in rats and pigs and was proven safe in phase I/II clinical studies. To enable the spatio-temporal presentation of regeneration factors, the alginate was modified with sulfate groups to mimic the binding of heparin-binding proteins to heparin/heparan-sulfate. When combining alginate-sulfate with the *in-situ* formed alginate hydrogel, multiple factor delivery was prolonged in ischemic tissues and enabled regeneration and cardiac repair after MI. The chapter emphasizes the increasing important role of biomaterials in various therapeutic strategies aimed at cardiac regeneration.

Introduction

Heart failure is the final consequence of a various set of insults to the myocardium, among which myocardial infarction (MI) is the most common cause. The limited ability of adult cardiomyocytes to replicate in response to injury contributes to the progressive damage following MI, causing the ventricular wall to thin and weaken.

S. Cohen (✉) · E. Ruvinov
Avram and Stella Goldstein-Goren Department of Biotechnology Engineering, Ben-Gurion University of the Negev, POB 653, 84105 Beer-Sheva, Israel
e-mail: scohen@bgu.ac.il

J. Hescheler, E. Hofer (eds.), *Adult and Pluripotent Stem Cells*,
DOI 10.1007/978-94-017-8657-7_6, © Springer Science+Business Media Dordrecht 2014

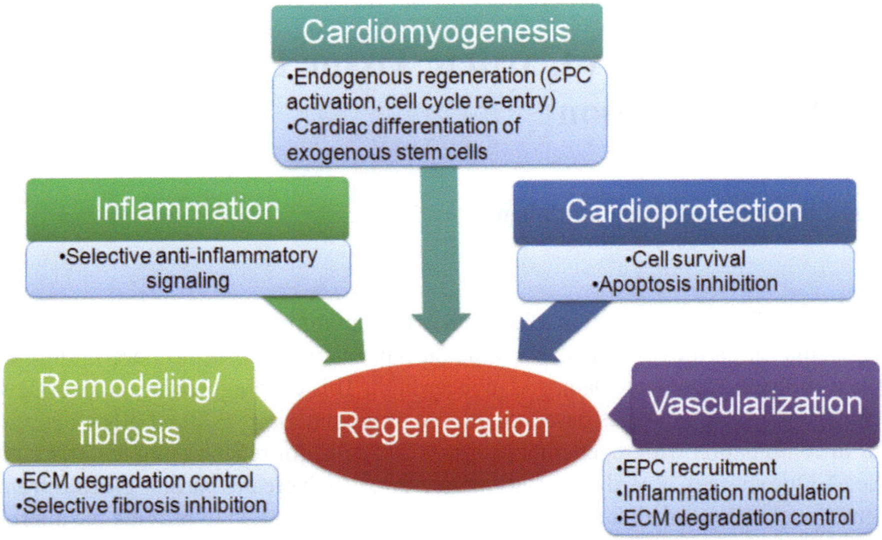

Fig. 6.1 Targeted therapeutic strategies for myocardial regeneration. (Reprinted with permission from Ruvinov et al. 2011a)

The progressive infarct expansion and left ventricular (LV) remodeling may lead to congestive heart failure (CHF). Currently, the wide-spread clinical remedies for CHF patients include pharmacological therapies; total heart transplantation; and LV assist devices, all of which do not lead to myocardial regeneration.

Over the last decade, new strategies for inducing cardiac muscle regeneration and repair have emerged and have shown great promise as potential therapeutic treatments of MI. The various strategies employ biomaterials, growth factors, and cells and aim to target major MI-associated processes (Fig. 6.1): (1) cardioprotection; (2) inflammation; (3) extracellular matrix (ECM) remodeling and fibrosis; (4) angiogenesis induction; and (5) cardiomyogenesis (Ruvinov et al. 2008).

This chapter presents an overview of strategies that are based on the employment of biomaterials, as stand-alone therapy or in combination with growth factors and cells, for repairing and regenerating infarcted hearts.

Cardiac Tissue Engineering

Cardiac tissue engineering aims to restore myocardial tissue loss after MI. This can be achieved by introducing to the infarct a fully-developed tissue graft prepared *in vitro* or by transplanting a cell-seeded scaffold and allowing the cells to organize into a matured tissue *in vivo*. To enable full replacement of the infarct, the engineered heart muscle should be thick and have to display the functional and morphological properties of the native cardiac muscle. Once integrated into the heart, the cardiac patch must develop systolic force, withstand diastolic load with appropriate compliance, and form an electrical and functional syncytium with the host myocardium (Zimmermann et al. 2006a).

In both the *in vitro* and *in vivo* tissue engineering strategies, the generation of the cardiac tissue microenvironment is required for the development of a functional cardiac tissue. Thus, biomaterials used for cardiac tissue engineering should mimic in their final structure and function, the native cardiac ECM. Although, the cardiac ECM composition is complex and hard to fully recreate, various types of natural and synthetic polymers have been shown to mimic (to various degrees) several important ECM properties, such as chemistry and hydrogel formation and degradability, thus they potentially have the ability to induce cardiac tissue regeneration.

Herein, we present the most applied strategies for fabrication of the cardiac patch, and describe recent advancements in biomaterial and bioreactor designs aimed to induce cardiac tissue regeneration, *in vitro*.

Strategies for Reconstructing the Cardiac Patch

Three main strategies for cardiac patch reconstruction have been developed over the last decade or so and they are based on the use of hydrogels, polymer layers and pre-formed macro-porous scaffolds (Fig. 6.2).

Cardiac cell entrapment in hydrogels In this strategy, cardiac cells are mixed with a liquid form of the biomaterial, followed by its solidification to create a mixed 3D cell hydrogel. This strategy has been developed by Eschenhagen and collaborators, who cultured embryonic cardiac myocytes in a collagen matrix to produce a coherently contracting 3D Engineered Heart Tissue (EHT) (Eschenhagen et al. 1997). Later, the same group improved the patch by suspending cardiomyocytes from neonatal rats in a mixture of collagen I and Matrigel and casting the solidified matrix into circular molds (Zimmermann et al. 2002). The construct was then subjected to a mechanical stretch, stimulating the tissue to differentiate and mature. The result was a ring-shape EHT that displayed important hallmarks of differentiated myocardium, such as a striated shape and an electrical response to a β-agonist (Zimmermann et al. 2002). Implantation of this patch into infarcted hearts showed undelayed electrical coupling to the native myocardium without arrhythmias, prevented LV dilation, induced systolic wall thickening and improved cardiac function (Zimmermann et al. 2006b). The EHT construct has also been found to be a valuable tool for preclinical toxicology (Schaaf et al. 2011).

Recently, Dahlmann et al. developed a fully defined *in situ* hydrogelation system based on alginate and hyaluronan (HA), in which their aldehyde and hydrazide-derivatives enable covalent hydrazone cross-linking of polysaccharides in the presence of viable myocytes. By varying the degrees of derivatization, concentrations and composition of blends in a modular system, the mechanophysical properties of the resulting hydrogels are easily adjustable. The hydrogel allowed for the generation of contractile bioartificial cardiac tissue from CM-enriched neonatal rat heart

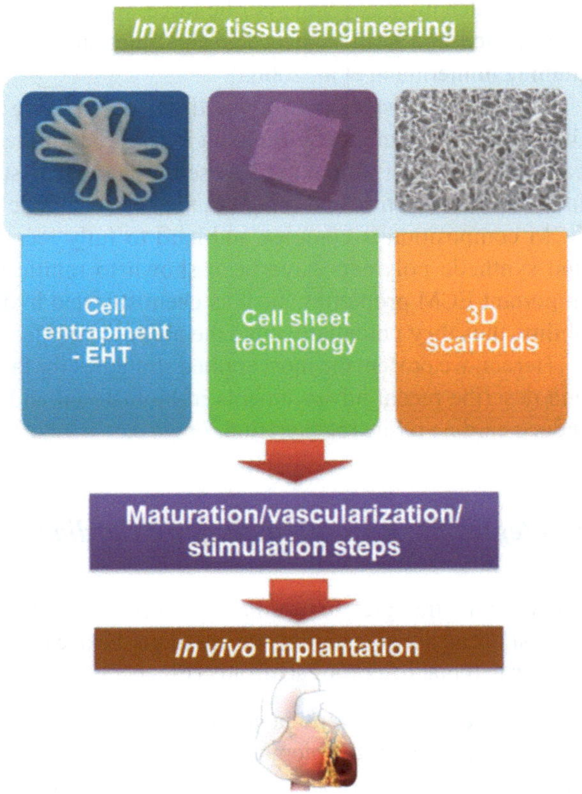

Fig. 6.2 Strategies for reconstructing the cardiac patch

cells, which resembles native myocardium. A combination of HA and highly purified human collagen I led to significantly increased active contraction force compared to collagen only (Dahlmann et al. 2013). The importance of this work is the attempt to create a fully defined matrix system composed of ECM protein and polysaccharide components, to significantly improve its clinical applicability.

Although the cardiac cell entrapment in hydrogels has led to the formation of beating cardiac tissue, the strategy in its present form could not promote thick tissue assembly (i.e. over 100 μm in thickness), mainly due to mass transfer limitations to the center of the construct.

Cell sheet-based cardiac tissue engineering Shimizu and co-workers developed a novel approach in which cardiac myocyte monolayers are assembled to form multilayered heart muscle constructs (Shimizu et al. 2002; Haraguchi et al. 2011). The cardiomyocyte layers were initially formed by cultivating isolated cardiac cells on cell culture surfaces, grafted with the temperature-responsive polymer, poly (*N*-isopropylacrylamide) (PIPAAm). Confluent cell monolayers could be detached from the surface as a cell sheet simply by reducing the temperature, without any enzymatic treatments. The released cell layers (2D) were overlaid one on top of the other, forming a multi-layer 3D tissue that began to pulsate simultaneously (Shimizu et al. 2002). After implantation on infarcted myocardium, gap junction formation and bi-directional smooth action potential propagation between the host heart and the grafted cardiac sheets were observed, without evidence of arrhythmias, suggesting functional and electrical graft integration (Furuta et al. 2006; Sekine et al. 2006). Cell sheet transplantation improved cell survival, angiogenesis and cardiac function, compared to dissociated cell injection (Sekine et al. 2011).

Myocardial tissue grafts based on preformed implantable scaffolds In this approach, cardiac cells are seeded within pre-formed macro-porous scaffolds. The scaffolds are generally prepared by cross-linking (chemical or physical) of biomaterial solution into the desired shape, with a subsequent solidification step and/or drying/freeze-drying to create porous matrix. This platform offers important advantages over cell entrapment during hydrogel fabrication, such as the ability to design and control the porosity of the scaffold prior to cell seeding and reduced exposure of the seeded cells to stress during mixing and molding of the scaffold. Furthermore, the pre-designed scaffolds can direct the seeded cells to organize into the gross conformation of native cardiac tissue, for example by influencing cardiac cell alignment in the porous structure.

Morphologically, the scaffold should be porous with an interconnecting pore structure to enable accommodation of a large number of cells and their subsequent organization into a functioning tissue (Shapiro and Cohen 1997). Pore size should be at least 50 μm to allow high mass transport during *in vitro* culture and vascularization of the scaffold following transplantation, so as to supply the seeded cells with nutrients and to remove secretions (Zmora et al. 2002). At the same time, the scaffold should present the desired mechanical properties and enable its handling in culture and during transplantation.

We were among the first groups to report on a successful implantation of cardiac cell-seeded porous alginate scaffolds into infarcted hearts (Cohen and Leor 2004; Leor et al. 2000). We found that the seeded fetal rat cardiac cells retained viability within the scaffolds and within 24 h formed multicellular beating cell clusters (Dar et al. 2002). Following implantation of the cellular constructs into the infarcted myocardium, some of the cells appeared to differentiate into mature myocardial fibers. The graft and surrounding area were populated with a large number of newly-formed blood vessels, consequently leading to attenuation in LV dilatation and improved heart function (Fig. 6.3; Leor et al. 2000). The efficacy of this strategy was further proven with patches prepared from various materials, natural (i.e., collagen, gelatin) or synthetic (i.e., poly(2-hydroxyethyl methacrylate-co-methacrylic acid) emphasizing the potential of this approach for myocardial repair (Kutschka et al. 2006; Li et al. 1999; Radisic et al. 2004a; Madden et al. 2010; Ye and Black 2011).

Recently, the use of decellularized heart matrices as scaffolds for cardiac tissue engineering has been described. Ott et al. decellularized whole rat hearts and re-seeded them with cardiac or endothelial cells. These "artificial" hearts generated modest but detectable contractions and pump function (Ott et al. 2008). Other groups utilized cardiac ECM matrices from pigs (Wainwright et al. 2009; Eitan et al. 2009; Godier-Furnemont et al. 2011). Collectively, the use of decellularized

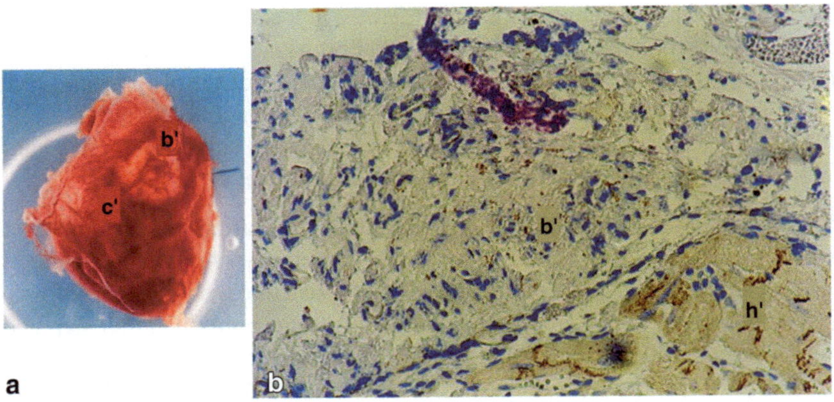

Fig. 6.3 Cardiac patch implantation on infarct. **a** Photograph of heart at week 9 after patch implantation. Visual examination revealed intense neovascularization into implanted patch (**a**-b′). Note a coronary branch (**a**-c′) that supplies and covers the patch with extensive network of vessels. **b** Microscopic image of integrated patch, immuno-stained for connexin-43 (Cx-43) (brown). Cx-43 was localized in normal parallel arrangements in host myocardium (**b**-h′) and randomly oriented in the patch (**b**-b′) (original magnification × 200). (Reprinted with permission from Leor et al. 2000)

matrices from natural ECM sources represents an attractive approach, as in this way the properties of the resulting scaffold can be directly matched to the target tissue, including the heart. However, several technical issues are still to be resolved, such as the need for effective decellurization protocols, possible immunogenicity, preservation/storage etc.

Bio-inspired Materials for Cardiac Tissue Engineering

Along with the use of ECM-extracted biological materials as scaffolds, major efforts have been invested in the intelligent design of novel materials bio-inspired by the ECM. These materials were created to provide the appropriate interface for cell interactions, including enhanced cell adhesion, growth, migration, and differentiation.

One strategy applied to achieve such appropriate material surfaces has been by attaching ECM-derived peptides. The most commonly used peptide is the sequence Arg-Gly-Asp (RGD), the signaling domain derived from fibronectin and laminin (Shin et al. 2003). RGD peptide mediates cell adhesion and signaling via binding of various ECM proteins to integrin receptors on the cell surface (Rosso et al. 2004).

Our group has covalently-attached RGD peptide to alginate, an algae-derived polysaccharide (Shachar et al. 2011; Tsur-Gang et al. 2009). Since alginate is inert and resistant to cell adhesion, it can be used as a "blank canvas" to investigate specific cell-matrix interactions by attaching RGD as well as other biological cues. Immobilization of RGD peptide into macroporous alginate scaffolds has proved to be a key parameter in cardiac tissue engineering, contributing to the formation of functional cardiac muscle tissue and to better preservation of the regenerated tissue culture (Shachar et al. 2011). More recently, we showed that the integration of an additional molecular signal provided by heparin-binding peptides (HBPs) promotes the formation of improved cardiac tissue *in vitro* (Sapir et al. 2011). HBPs, with the sequence XBBXBX and XBBBXXBX, where B is a basic amino acid and X is a hydropathic amino acid, have been shown to bind cell surfaces via the cellular syndecans, which bear 3–7 covalently-attached heparin/heparan-sulfate chains (Cardin and Weintraub 1989). Thus, we compared cardiac tissue engineering of neonatal rat cardiac cells cultivated within macro-porous alginate scaffolds bearing the two peptides, RGD and HBP, to those formed in a single peptide-attached or in unmodified alginate scaffolds (Sapir et al. 2011). Peptide modification did not affect scaffold architecture, pore size, and stiffness. Examination of cardiac cell morphology 14 days after seeding in the different scaffolds revealed characteristic striated fiber organization mainly in the RGD/HBP-attached cell constructs, while in HBP-attached and unmodified alginate cell constructs no such structures were observed. By histology, the 14-day constructs in HBP/RGD-attached scaffolds presented an isotropic arrangement of the fibers in the form of consistent tissue, while in the RGD-attached scaffold no such arrangement was seen (Fig. 6.4). Western blot analysis of selected cardiac markers (sarcomeric α-actinin, N-cadherin and connexin-43 (Cx-43)) showed preservation of these proteins and an increase in expression level of Cx-43 with time in the HBP/RGD-attached scaffolds, further supporting the notion of contractile muscle formation and tissue maturation (Fig. 6.4; Sapir et al. 2011). Collectively, the results of this study emphasize the importance of integration of multiple and complementary signals into 3D matrices for the recreation of the cardiac microenvironment and successful formation of an engineered cardiac tissue *in vitro*.

Perfusion Bioreactors for in vitro Cardiac Tissue Engineering

In early studies, cells were seeded onto scaffolds and cultivated in dishes, spinner flasks or rotating vessels. In all of these systems, the dissolved oxygen in culture medium was transported to the seeded cells by molecular diffusion. Since human cardiac muscle is ~1 cm thick, oxygen diffusion alone can support only four to seven cell layers, that is, a 100-μm thick layer of viable and compact tissue (Radisic et al. 2007; Radisic et al. 2008; Shachar and Cohen 2003). Perfusion bioreactors with interstitial medium flow have been designed to overcome these limitations and to enable *in vitro* tissue engineering.Perfusion bioreactors operate by forcing the culture medium through a fixed porous 3D cell construct, thus providing a controllable, mechanically active environment for all seeded cells. Cultivation of engineered tissues in perfusion bioreactors improved the construct size, cellularity, and molecular composition, and enabled studies such as the effect of (i) mass transport of oxygen

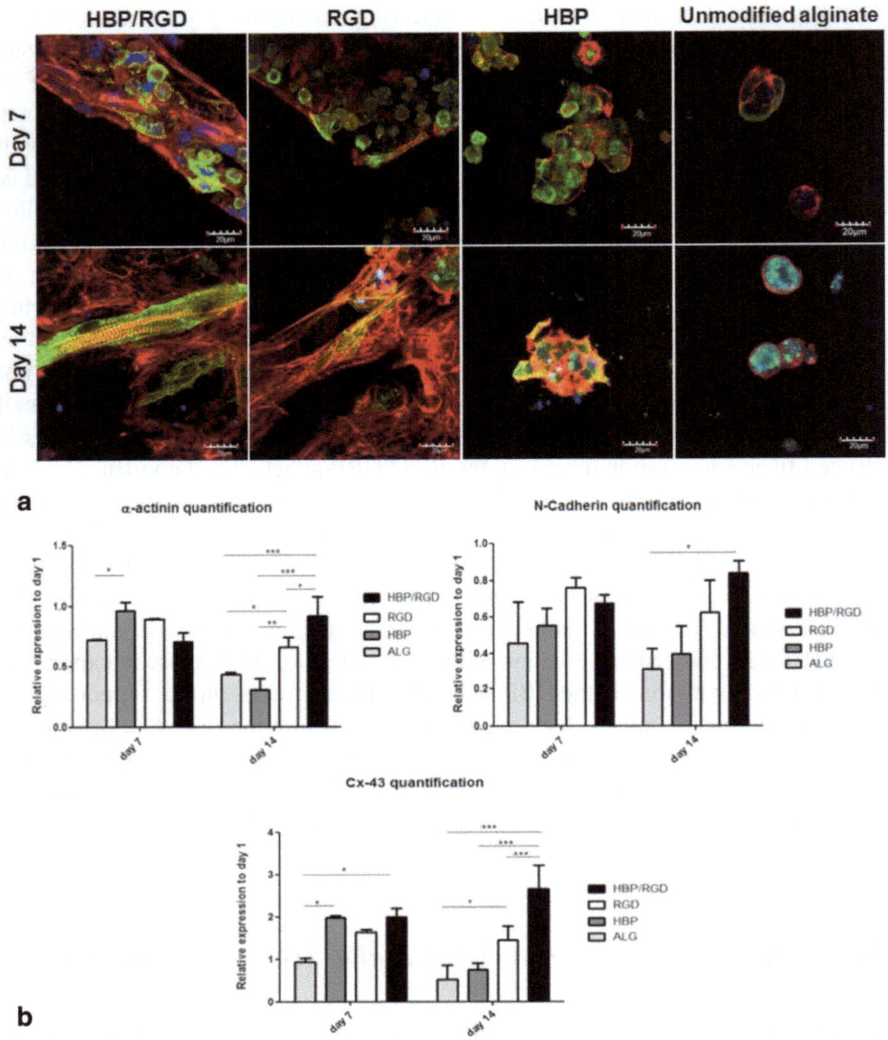

Fig. 6.4 Integration of multiple cell-matrix interactions into alginate scaffolds promotes cardiac tissue organization. (**a**) Cardiac cell organization, on days 7 and 14 post cell seeding. Cardiomyocytes are positive for α -actinin (*green*), while all cells are stained for F-actin (*red*) and nuclei (*blue*) (bar: 20 μm). (**b**) Protein expression of representative cardiac cell markers. Asterisks denote significant difference (by 2-way ANOVA), $*p<0.05$, $**p<0.01$ and $***p<0.005$ (Bonferroni's pos-hoc test was used for comparison between the groups). (Reprinted with permission from Sapir et al. 2011)

and growth factors, (ii) hydrodynamic conditions, and (iii) physical stimuli, such as dynamic compression or cyclic stretch, on the developing tissue (Freed et al. 2006).

The initial perfusion bioreactors were commercially-available cartridges of 13 mm filter holders, connected to a peristaltic pump. A stainless steel screen was placed at the cartridge inlet to disperse the medium flow over the construct surface, while a nylon mesh was used to fix the downstream side of the construct (Carrier et al. 2002). Cultivation under pulsatile flow in such system enhanced the contractile properties and improved cell viability in the constructs, yet these features were confined mainly to construct periphery (Radisic et al. 2004b; Brown et al. 2008). Probably, the non-homogenous tissue formation resulted from the use of construct mesh holders with only 60–70 % porous structure, resulting in a non-homogeneous milieu within the system.

Our group designed a perfusion bioreactor system with two unique cell construct-fixing nets, with an open area of 95.8 % (Dvir et al. 2006). The net architecture is an assembly of equally spaced, 380-square-based micro-pyramidal structures, with their heads pointing toward and holding the multiple scaffolds. Four round openings are located at the corners of the square pyramide, enabling medium perfusion from the small holes toward the pyramid head, at an angle of 60°, thus maintaining the fluid flow direction. Using this net architecture, 99.88 % of the cell construct volume is perfused by the culture medium. In addition, to impose equal shear stress along the bioreactor cross-section, we inserted a micromesh designed to disrupt the perfusing media for a split second before entering the cell constructs (Fig. 6.5). The mesh transforms the developed laminar fluid flow velocity profile into an undeveloped one, having equal velocity vectors along the bioreactor cross-section and subjecting the cellular constructs to an equal shear stress. By employing this advanced perfusion bioreactor, the cardiac cell constructs maintained viability of almost 100 % of the seeded cells after a 7-day cultivation period, while less than 60 % of the cells in static cultures were viable at the end of this period. Moreover, a homogenous thick (> 500 μm) cardiac tissue was generated, composed of elongated and aligned cells with a massive striation. Ultrastructural morphology analyses revealed organized sarcomeres, defined Z-lines and intercalated disks, resembling the native heart tissue (Dvir et al. 2006).

Our group has further investigated the effects of the unique microenvironment provided by the perfusion bioreactor described above, on cardiac tissue assembly. For the first time, the various molecular and cellular events taking place in the cell construct, from the cell signaling level up to the ultrastructural phenotype of the regenerated cardiac muscle tissue, have been elucidated (Dvir et al. 2007). We demonstrated that immediately after operating the bioreactor at a shear stress of 0.6 dynes/cm^2, the extracellular signal-regulated kinase 1/2 signal transduction pathway is activated, inducing the synthesis of cardiac contractile proteins (Troponin T and sarcomeric α-actinin) and cell-cell interaction proteins (Cx-43 and N-cadherin). Furthermore, we showed that prolonged cultivation under this stimulation regime promoted the assembly of a thick cardiac tissue, in many respects resembling the native adult heart (Dvir et al. 2007).

In addition to mechanical stimulation, it has been realized that the electrical stimulation of cardiac cell construct *in vitro*, which mimics electrical signals present in native heart tissue, promotes the development of conductive and contractile properties characteristic of cardiac tissue, including cell alignment and coupling, increased amplitude of synchronous construct contractions and a remarkable level of ultrastructural organization (Radisic et al. 2004a; Tandon et al. 2009). We combined

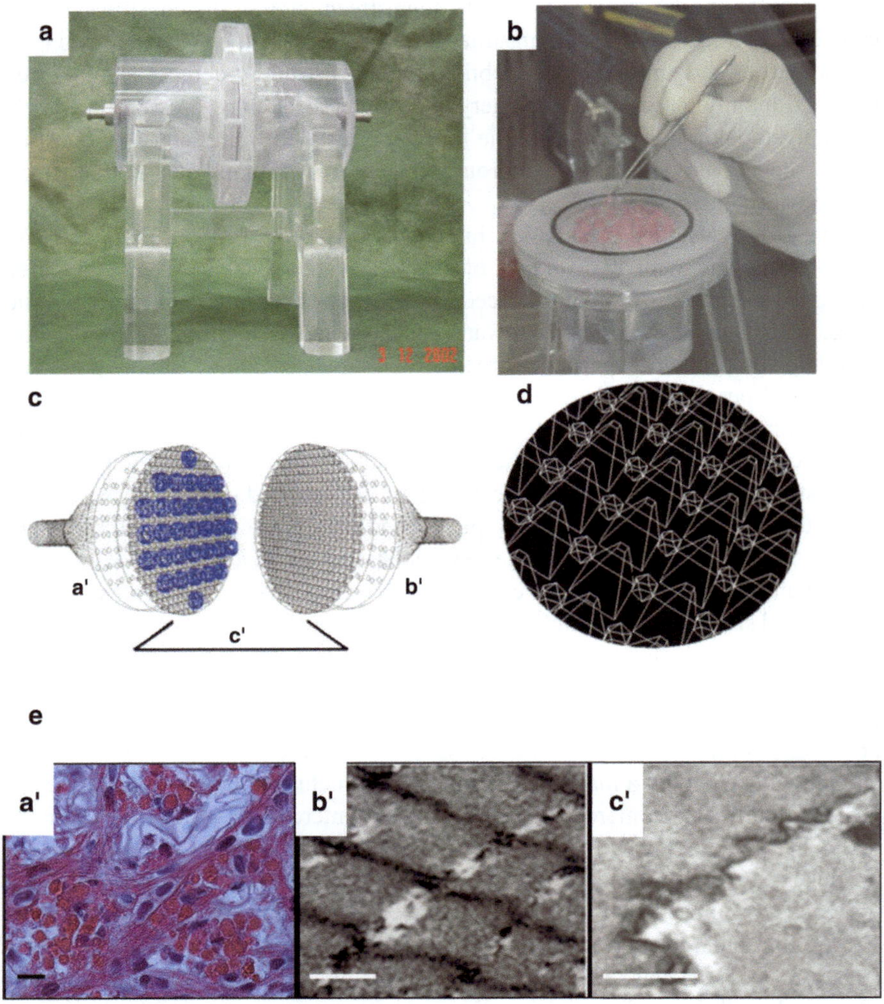

Fig. 6.5 Perfusion bioreactor system. **a** Overall view of the bioreactor. **b** Multiple cell constructs are placed on the fixing net. **c** Schematic representation of the bioreactor vessel. Two identical halves of the bioreactor (**c**-a′, **c**-b′) function as the inlet and the outlet of the cell construct compartment (**c**-c′). Multiple cell constructs (*blue*) are fixed between two open pore nets. The angles of both halves were designed to be 60° to prevent the breakdown of the incoming flow. **d** Architecture of the fixing nets. The nets are comprised of 380-square-based pyramidal structures with their heads pointing towards and holding the cell constructs. **e** Histo and ultrastructural morphology of the engineered tissue. **e**-a′ H&E staining reveals massive striation and elongation of the cultured cells. **e**-b′ TEM images show defined Z-lines and multiple high-ordered sarcomeres and **e**-c′ intercalated disks between adjacent cardiomyocytes. Scale bars: a', 20 μm; b', 1 μm; c', 0.5 μm. (Dvir et al. 2006, 2007).

electrical field stimulation within the perfusion bioreactor and showed that cultivation of cardiac cell constructs for only 4 days under perfusion and continuous electrical stimuli promoted cell elongation and striation in the cell constructs and enhanced the expression level of the gap junction protein Cx-43 (Barash et al. 2010).

The implementation of perfusion bioreactors in tissue engineering has greatly advanced the field, leading to the construction of functioning tissues with thickness much greater than seen before.

Vascularization Approaches for Cardiac Patches

Vascularization of the cardiac patch prior to its implantation on the infarcted myocardium is a great challenge and the realization of this strategy in clinical protocols depends on achieving this goal. Without immediate vascularization after implantation, the thick cardiac patches do not survive in the anoxic infarcted zone. Several strategies have been employed to induce vascularization of engineered cardiac patches. We categorize these strategies as direct and indirect ones. In direct approaches, the cell constructs include endothelial cells in addition to the functional cells. Indirect approaches include for example scaffold designs with discrete compartments for specific cell types and/or optimized for better vascularization upon implantation (Maidhof et al. 2010; Madden et al. 2010). Other indirect approaches include incorporation of proangiogenic growth factors (such as VEGF, PDGF-BB and bFGF), either covalently immobilized to the scaffold or in the form of controlled delivery systems (Chiu et al. 2010; Miyagi et al. 2011; Saik et al. 2011; Perets et al. 2003; Gao et al. 2011).

The most studied direct approach for *in vitro* patch pre-vascularization is the formation of a microvascular network within the engineered construct by co-culture of endothelial cells (ECs), embryonic fibroblasts and cardiomyocytes (CMs). Caspi et al. reported the formation of synchronously contracting engineering human cardiac tissue in a co-culture of cardiomyocytes (CM), ECs and embryonic fibroblasts (EmFs), derived from human embryonic stem cells (hESC). The construct revealed the formation of endothelial vessel networks, stabilized by the presence of mural cells originating from the EmFs (Caspi et al. 2007). In a similar approach, Lesman et al. developed an engineered human cardiac tissue by co-culture of hESC-derived CMs, EmFs and human umbilical vein ECs (HUVECs). Transplantation of these constructs onto healthy rat hearts resulted in formation of both donor (human) and host (rat)-derived vasculature within the constructs, associated with functional integration of human-derived vessels with host coronary vasculature (Lesman et al. 2009). Murry and co-workers used co-cultures of hESC- or induced pluripotent stem cell (iPSC)-derived CMs with HUVECs to engineer human cardiac muscle. Addition of HUVECs increased CM proliferation, and the further addition of stromal supporting cells (mouse embryonic fibroblasts and human marrow stromal cells) enhanced formation of vessel-like structures by ~10-fold. Subjection of the co-culture to a continuous mechanical load yielded a significantly more aligned tissue with good contractility. When transplanted onto healthy hearts of athymic rats, the engineered tissue formed stable grafts revealing the presence of human microvessels that were perfused by the host coronary circulation (Tulloch et al. 2011).

A different approach to vascularize a tissue is by using the body as a continuous providing source of ECs and pericytes. The body is essentially used as a bioreactor, serving as a humidified, temperature-controlled incubator. This bioreactor can promote essential processes during tissue development, such as cell proliferation and differentiation. Lee and colleagues quantitatively compared the omentum, mesentery and subcutaneous space as possible sites for the engraftment of an engineering tissue. According to their results, the highly vascularized omentum is the preferred tissue for engraftment and vascularization due to high secretion of growth and angiogenic factors, which promote angiogenesis and cell survival in the graft (Lee et al. 2003; Zhang et al. 1997). Suzuki et al. showed that the omentum was able to maintain the viability and angiogenesis of transplanted myocardial cell sheets (Suzuki et al. 2009).

We sought to use the rich blood vessel environment of the omentum to promote vascularization of a cardiac patch prior to its implantation on the infarct (Dvir et al. 2009). Scaffolds prepared from affinity-binding alginate (to control growth factor release and presentation) loaded with a cocktail of pro-survival and angiogenic factors (insulin-like growth factor-1 (IGF-1), stromal cell-derived factor-1 (SDF-1) and VEGF) were seeded with rat neonatal cardiomyocytes, and then transplanted onto the omentum to achieve host-induced vascularization of the engineered tissue. After 7 days on the omentum, the vascularized omentum-generated patches were implanted on a 7-day old scar tissue in rat hearts. Four weeks later, the omentum-generated patches were shown to structurally and electrically integrate into the scar tissue, attenuate LV dilatation and remodeling, and improve cardiac function (Fig. 6.6; Dvir et al. 2009).

Interestingly, similar beneficial results were obtained when the omentum-generated patch was constructed from an affinity-binding scaffold that was supplemented with only the mixture of pro-survival and angiogenic factors without seeded cardiac cells (Dvir et al. 2009). After regeneration on the omentum, these acellular constructs showed similar cell penetration and consistency as the cardiac-seeded constructs, which may explain their beneficial effects on infarct repair.

Acellular Biomaterials for Inducing Cardiac Regeneration

In parallel to investigating cellular-based strategies to treat MI, our group has been researching, over the last decade or so, a-cellular approaches based on injectable alginate-based biomaterials.

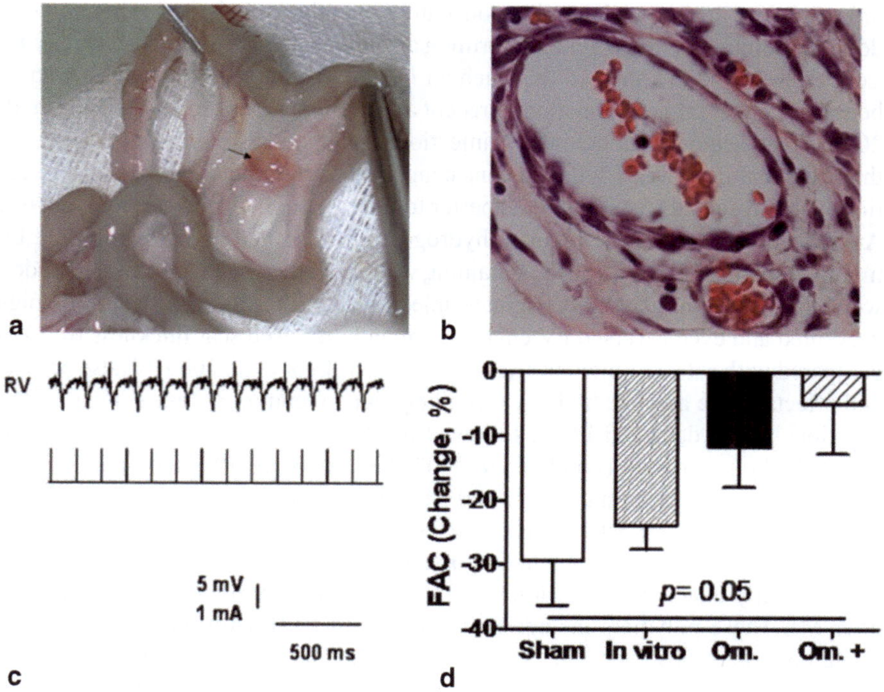

Fig. 6.6 Prevascularization of a cardiac patch created in affinity-binding alginate scaffolds on the omentum improves its therapeutic outcome. **a** The cardiac patch is transplanted on the omentum for 7 days to promote its vasculogenesis. **b** Functional blood vessels within the patch. **c** Twenty eight days post transplantation of the vascularized patch on the infarcted heart, stimulation of the patch was able to trigger synchronized beating of the healthy right ventricle. **d** Fractional area change (FAC) of infarcted hearts after treatment with a stitch (sham), implantation of *in vitro*-grown patch, empty patch grown on the omentum (Om), or omentum-generated cardiac patch (Om+) (Dvir et al. 2009).

Injectable Alginate Biomaterial as ECM Replacement

The ECM plays a critical role in maintenance of the integrity and function of the heart. Dynamic changes and excessive damage to cardiac ECM following MI have a detrimental effect on both systolic and diastolic function of the heart (Akhyari et al. 2008; Dobaczewski et al. 2010; Jugdutt 2003; Spinale 2007; Iraqi et al. 2009). Thus, we hypothesized that injection of biomaterials into the infarct could replace damaged ECM, preventing the progression of heart failure. The application of injectable, rather than solid forms (e.g., scaffolds) of biomaterials for this goal is less invasive, and therefore more clinically appealing.

To address this aim, our group developed an injectable alginate biomaterial, which due to its non-thrombogenic nature can be delivered by intracoronary injection as a solution. When delivered to the infarct-related arteries, the alginate material was able to transverse the leaky vessels to the infarct. Then, and only at the infarct, due to the high calcium ion concentration and relatively low water content, does this solution undergo gelation forming an alginate hydrogel (Landa et al. 2008; Leor et al. 2009). We showed that such an *in situ* forming alginate hydrogel has a beneficial effect on heart function in recent and chronic models of MI (Landa et al. 2008). In an acute MI model in rats, injection of alginate resulted in increased scar thickness and improved LV dimensions compared with saline, 2 months post injection. These results were similar or superior to injection of neonatal cardiomyocytes. At 6 weeks post injection, alginate hydrogel was degraded and was replaced by myofibroblast infiltration, thus maintaining wall thickness. In a large animal model, we showed that intracoronary alginate injection into the infarcted hearts of pigs prevented and even reversed LV enlargement and increased scar thickness by 53 % compared with saline, 2 months after the injection (Leor et al. 2009). These beneficial effects of the alginate hydrogel are likely due to temporary replacement of the functions of the damaged ECM, followed by increased myofibroblast infiltration during hydrogel erosion (Landa et al. 2008; Leor et al. 2009). These results have recently led to a first-in-man safety and feasibility study of intracoronary injection of alginate in acute MI patients (BioLineRx Ltd. 2007).

Various other natural or synthetic polymers can also form hydrogels, and after being directly injected into the infarcted heart have shown to improve the outcome after MI. These polymers include collagen, fibrin, PNIPAAm and methacrylated hyaluronic acid; however, due to their thrombogenic nature, they cannot be administered systemically (e.g., intracoronary administration) and are delivered by a direct myocardial injection (Dai et al. 2005; Christman et al. 2004a, b; Masuda et al. 2008; Shimizu et al. 2003; Wang et al. 2009; Yu et al. 2009; Mukherjee et al. 2008; Ifkovits et al. 2010).

Evolution of the Active Biomaterial Approach for Myocardial Regeneration

The functional improvement observed after treatment with acellular biomaterials has not been associated with actual tissue regeneration, meaning no new contractile units were added to the infarct. Instead, the beneficial effects of the *in situ* forming hydrogels are most likely related to their capability of increasing scar thickness and stabilizing the early infarct, by providing scaffolding and critical physical support to the LV wall, as well as replacement for the damaged ECM.

To induce active myocardial regeneration and introduce new viable beating tissue, combinations of regenerating signal factors are required. Biomaterials can provide a temporary matrix support and simultaneously direct the formation of a functional tissue by providing a temporary depot for the sustained delivery of bioactive molecules, with spatial and controlled distribution of the desired agents (Christman and Lee 2006; Davis et al. 2005).

The concept of affinity-binding alginate biomaterial To enable the precise control over factor release and to allow the release of combinations of growth factors, we designed an alginate biomaterial with affinity binding sites for heparin-binding proteins, by sulfation of the uronic acid monomers in alginate (Freeman et al. 2008). Surface Plasmon Resonance (SPR) analysis revealed the strong binding of various heparin-binding proteins to alginate-sulfate (Freeman et al. 2008; Freeman and Cohen 2009). The binding to alginate-sulfate mimics an example from nature, where many growth factors, chemokines, and cell adhesion molecules, collectively known as heparin-binding proteins, bind heparin and heparan sulfate via high affinity, specific electrostatic interactions with the low- and high-sulfated sequences in these glycosaminoglycans (GAG) (Shriver et al. 2002). In this aspect, heparan sulfate GAGs play an important role in sequestering and storage of the proteins, and also participate in the formation of an active signaling complex with a respective cell surface receptor. The development of alginate-sulfate has enabled to mimic these properties of heparan sulfate/heparin with an added advantage of providing a stable and prolonged delivery system. Alginate and alginate-sulfate are not recognized by enzymes in the human body, thus the erosion of the hydrogels is slow, and this adds another barrier for protein release.

Combination of alginate-sulfate with unmodified alginate represents a unique type of a novel *affinity-binding* alginate biomaterial, which is capable of controlling the presentation and delivery of multiple proteins, while retaining the properties and characteristics of the alginate hydrogel as an ECM-replacing material. Macroporous scaffolds prepared from affinity-binding alginate were able to sequentially release the combination of three known angiogenic factors (vascular endothelial growth factor (VEGF), platelet-derived growth factor-BB (PDGF-BB) and transforming growth factor-β1 (TGF-β1)) in an order predicted from their respective equilibrium binding constants to alginate-sulfate. This order of release coincides with their participation in blood vessel formation and maturation. Subcutaneous implantation of the triple factor-loaded affinity-binding alginate scaffolds in rats increased blood vessel density and maturation in the construct and surroundings, 3 months after implantation, compared to unmodified scaffolds (Freeman and Cohen 2009). The affinity-binding alginate scaffolds were also used for the creation of a vascularized cardiac patch, as described in a previous chapter (Dvir et al. 2009).

Injectable affinity-binding alginate biomaterial and factor delivery To enhance the clinical applicability of the concept of affinity-binding alginate for myocardial repair, the system was produced in an injectable form. We achieved this by mixing the alginate-sulfate/growth factor bioconjugates with partially cross-linked alginate solution. This injectable affinity-binding alginate system can be easily delivered into the infarcted heart where it creates a hydrogel capable of a sustained delivery of multiple growth factors (Fig. 6.7). We hypothesized that such an affinity-binding alginate hydrogel would have two main functions: (1) provide temporary tissue support and replace damaged ECM, together with temporary mechanical stabilization of the infarct, as previously shown for this unique alginate biomaterial; and (2) control the release and presentation of multiple growth factors.

We chose IGF-1 and hepatocyte growth factor (HGF) as the bioactive components of the injectable affinity-binding alginate system, due to their established

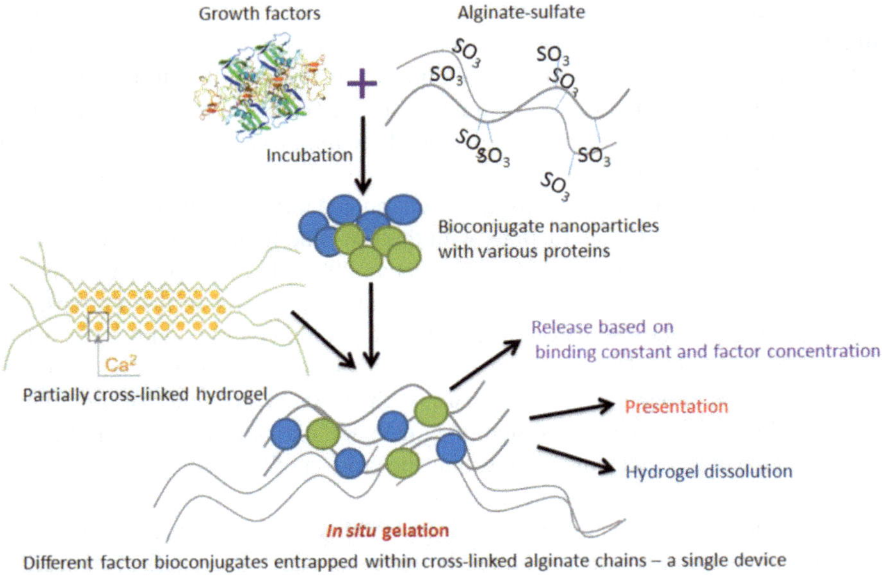

Fig. 6.7 The concept of injectable affinity-binding alginate biomaterial, and mode of protein release and action. (Reprinted with permission from Ruvinov et al. 2011a)

and complimentary beneficial effects on infarcted myocardium (Conti et al. 2004; Nakamura et al. 2000; Vandervelde et al. 2005; Hausenloy and Yellon 2009). The affinity-binding alginate hydrogel was able to release biologically active proteins in a sequential order, IGF-1 followed by HGF (Ruvinov et al. 2011b). Importantly, the sequential delivery of IGF-1 and HGF is suited for the proper execution of the reparative processes in the infarcted myocardium, and for achieving a more favorable course of repair (Fig. 6.8). The faster released IGF-1 could provide an immediate strong pro-survival signal to rescue the functional myocardium and reduce cell apoptosis after the initial ischemic event (Suleiman et al. 2007; Li et al. 1997; Webster 2007). Processes required at later phases of repair, such as angiogenesis induction, more favorable ECM remodeling and fibrosis reduction, can be mediated by the slower, yet continuous, release of HGF (Tomita et al. 2003; Nakamura et al. 2005; Wang et al. 2004).

Increased retention of therapeutic proteins over time is one of the main attributes of a successful therapy. The infarct after the initial ischemic event represents a very hostile environment, where extensive protein degradation takes place as part of inflammation and ECM remodeling-induced enzymatic responses (Dobaczewski et al. 2010; Frantz et al. 2009). Thus, we chose an acute MI and immediate post-MI injection as a model for testing the efficacy and impact of our delivery system on protein retention. When delivered with the affinity-binding alginate, HGF retention and bioavailability in the myocardial tissue after acute MI were increased, in comparison to soluble HGF administered by bolus injection, which was rapidly eliminated from the infarct (Ruvinov et al. 2010; Fig. 6.9). The affinity-binding alginate

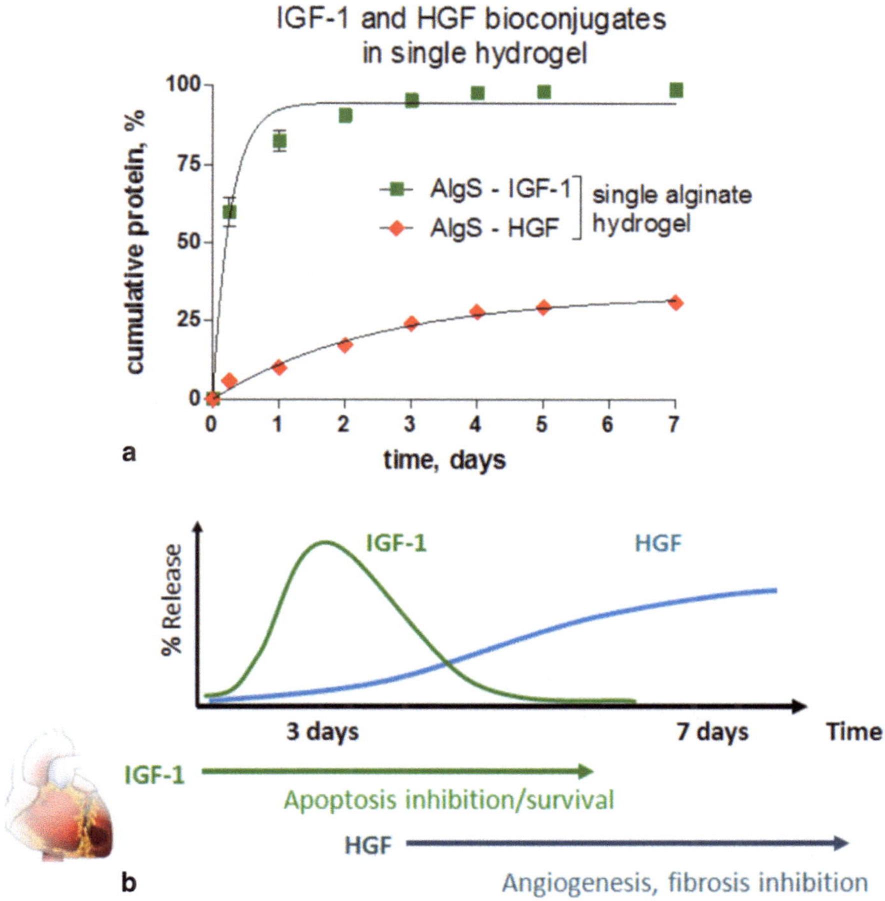

Fig. 6.8 A sequential release pattern of the factors, IGF-1 followed by HGF, from affinity-binding alginate hydrogel. **a.** *In vitro* release from dual factor-loaded system containing both protein bioconjugates. **b.** Correlation of growth factor release profile from affinity-binding alginate hydrogel with required reparative processes after MI. (Reprinted with permission from Ruvinov et al. 2011a)

was also able to translate the known angiogenic effect of HGF into an improved therapeutic outcome in a murine model of hindlimb ischemia, where controlled delivery of HGF by this system improved tissue blood perfusion and induced the formation of a mature blood vessel network (Fig. 6.9; Ruvinov et al. 2010).

Ultimately, we tested the therapeutic efficacy of multiple growth factor delivery by affinity-binding alginate hydrogel in an acute MI model. We found that the sequential delivery of IGF-1 and HGF from this system reduced scar fibrosis, increased scar thickness and prevented infarct expansion, 1 and 4 weeks after injection, compared to controls. Furthermore, angiogenesis was induced and cell apoptosis was reduced at the infarct (Fig. 6.9). These therapeutic effects were preserved for 8 weeks after injection (unpublished data). This treatment also exhibited signs

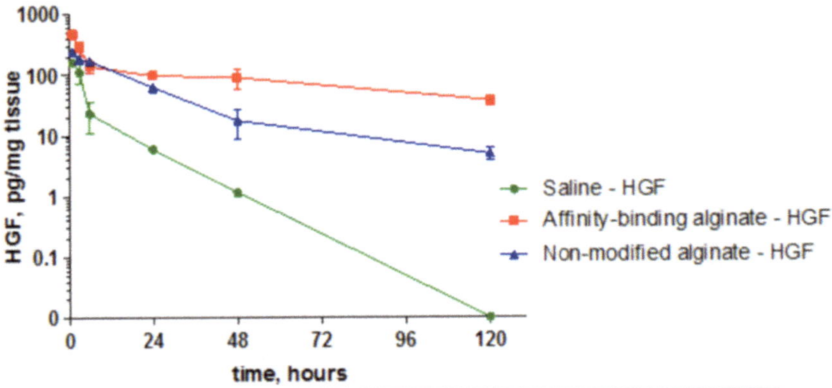

Parameter	Saline - HGF	Affinity-binding alg - HGF	Non-modified alg – HGF
AUC – measure of bioavailability, pg×h/mg tissue	1002	10322	4532

a

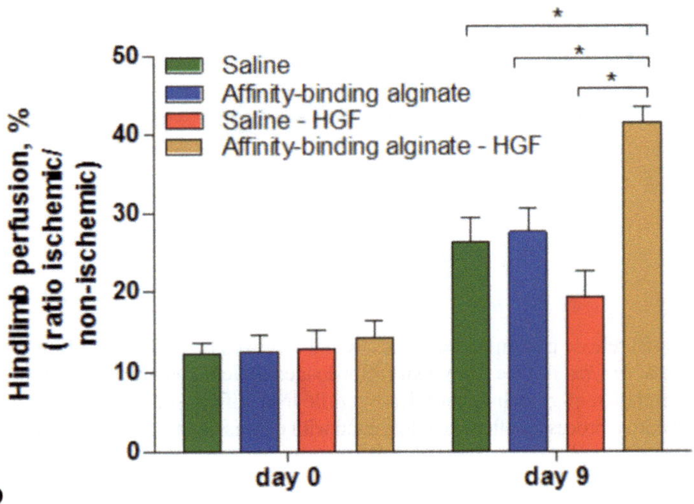

b

Fig. 6.9 Therapeutic efficacy of the dual factor delivery from the injectable affinity-binding alginate system. **a** Myocardial tissue retention of HGF in a rat acute MI model. *AUC* area under the curve. **b** Delivery of HGF by injectable affinity-binding alginate biomaterial improves limb perfusion in mice hindlimb ischemia model. $*p<0.05$. **c** The effects of sequential delivery of IGF-1/HGF by injectable affinity-binding alginate biomaterial on various aspects of tissue regeneration. Infarct expansion index was evaluated by Masson's trichrome staining. Blood vessels were identified by α-SMA staining. Apoptotic cells were identified by active caspase-3 staining. $*p<0.05$. **d** The effect of sequential IGF-1/HGF delivery on Ki67 and GATA-4 expression. **d**-a′ Quantitative analysis of Ki67-positive cardiomyocytes (based on morphological discrimination) in short-term (1-week) experiment. $*p<0.001$. **d**-b′ Incidence of GATA-4-positive cell clusters in different treatment groups. Incidence = (number of animals with positive staining/total number of animals in each group) × 100. (Reprinted with permission from Ruvinov et al. 2011a)

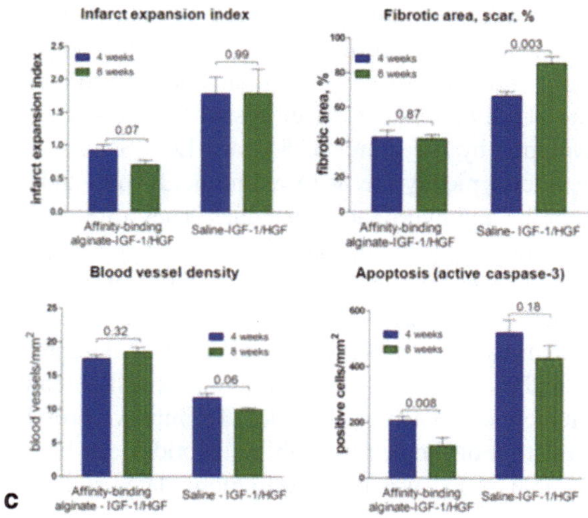

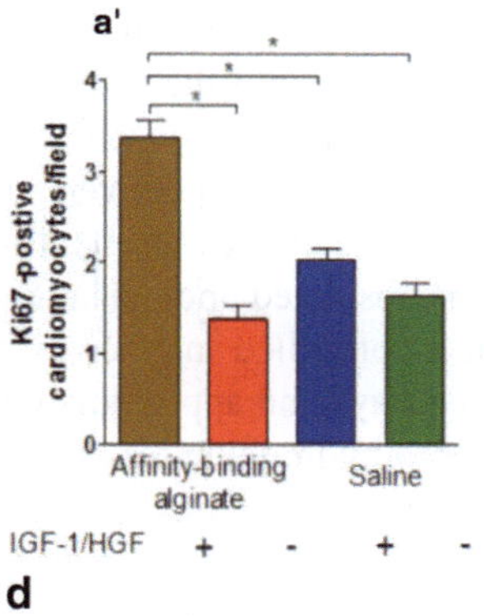

b'

GATA-4 incidence

	Affinity-binding alg-HGF+IGF-1	Affinity-binding alg	Saline-HGF+IGF-1	Saline
1 week	1/5 (20%)	0/5 (0%)	1/5 (20%)	0/5 (0%)
2 weeks	4/5 (80%)	0/4 (0%)	2/5 (40%)	0/4 (0%)

Fig. 6.9 (continued)

that could be associated with endogenous regeneration of the cardiac muscle (increased incidence of Ki67-postive cardiomyocytes and GATA-4-positive cell clusters) (Fig. 6.9; Ruvinov et al. 2011b). These results were deduced based on simple morphological discrimination and single marker staining. However, more extensive analyses (e.g., genetic mapping, simultaneous staining) are required to confirm the effects of sequential IGF-1/HGF delivery on endogenous myocardial regeneration.

The results observed after the treatment with the sequentially-delivered proteins from the affinity-binding alginate system collectively indicate that marked salvage of the functional tissue occurred due to the enhanced retention of the bioactive factors at the infarct and their sequential release pattern, mediated by the affinity-binding mechanism to alginate-sulfate. These effects, when accompanied by passive tissue support and mechanical stabilization of the infarct conferred by the alginate hydrogel itself, led to cardiac repair (Landa et al. 2008; Leor et al. 2009).

Other modalities of injectable biomaterial-based growth factor delivery and the advantages of the affinity-binding alginate Several polymeric systems have been applied for the delivery of various therapeutic proteins directly into infarcted hearts. One of the best characterized systems is the self-assembling peptide nanofibers developed by the group of Richard Lee (Segers and Lee 2007). Self-assembling peptides typically are 8–16 amino acids long and are composed of alternating hydrophilic and hydrophobic residues. They form stable β-sheets in water, and upon exposure to physiological salt concentration or pH they form a stable hydrogel of flexible nanofibers (7–20 nm in diameter) consisting of more than 99 % water (Segers and Lee 2007). Slow release of the proteins (e.g., PDGF-BB) from such a system has been achieved by the physical entrapment of the protein in the hydrogel and possibly by its adsorption on the self-assembling peptides by non-covalent interactions (Hsieh et al. 2006). To improve protein retention in the hydrogel, biotinylation of the self-assembling peptides can be performed. By this method, for example, IGF-1 was tethered to biotinylated self-assembling peptides (Davis et al. 2006). To improve protein stability and confer protection from proteolysis, Segers et al. genetically engineered a protease-resistant form of SDF-1 that was subsequently delivered by self-assembling nanofibers into the infarcted heart by intramyocardial injection (Segers et al. 2007). Recently, Radisic and colleagues developed a collagen-chitosan hydrogel capable of controlled delivery of thymosin-β4 (Tβ4). This hydrogel was composed of a 2:1 collagen to chitosan ratio. Tβ4 was mixed with the hydrogel solution and became encapsulated upon gelation. The hydrogel could deliver biologically active Tβ4 in a controlled manner over 28 days, with a burst release of 18–28 % over the initial 3 days and approximately zero order kinetics from day 3 to 28. The collagen–chitosan hydrogel had a porous structure and gelation time of 200–300 s. *In vivo,* Tβ4-encapsulated hydrogel was injected into the infarct region in rat MI model. After 3 weeks, there was reduced wall thinning and tissue loss, an increased blood vessel density after injection of Tβ4-encapsulated hydrogel, compared to hydrogel alone (Chiu et al. 2012).

In addition, multiple and/or sequential factor delivery has been described by several groups, applying different delivery systems or different combinations of polymers (Richardson et al. 2001; Lu et al. 2007). For example, Hao *et al* used a combination of partially oxidized alginates with low and high molecular weights to produce a hydrogel system that can sequentially deliver VEGF and PDGF-BB into the infarcted myocardium (Hao et al. 2007). This delivery pattern was achieved due to the different degradation rates of the partially oxidized alginates constituting the hydrogel.

Compared to the above modalities, the affinity-binding alginate system provides several potential advantages. First, the inclusion of the factors in our system does not involve covalent chemistry or non-specific binding of the factor to the biomaterial; instead it relies on mimicking the natural specific affinity of factors to the matrix. Second, the affinity-binding mechanism enables binding of multiple proteins and their release is well-controlled by the protein's equilibrium binding constant and its initial concentration. Third, the bioconjugation between the factor and alginate-sulfate results in the formation of nanoparticles so that the proteins are shielded from enzymatic proteolysis by an external hydrogel layer (Ruvinov et al. 2010, 2011b). Finally, due to the *in situ* hydrogel formation, the affinity-binding alginate-based release system may serve as an ECM replacement and provide temporal tissue support and an improved microenvironment for more effective self-repair and possible progenitor cell recruitment.

Clinical Status of Biomaterial Application for Myocardial Repair

An exciting development over the last decade is the beginning of clinical translation of cardiac patches and injectable materials.

In the MAGNUM (Myocardial Assistance by Grafting a New Bioartificial Upgraded Myocardium) nonrandomized controlled phase 1 clinical trial, bone marrow cells seeded in collagen type 1 patch were implanted onto 10 patients that had coronary artery bypass graft and were intramyocardially injected with suspensions of autologous bone marrow cells. In 10 other patients, the treatment was only cell suspension injection into the infarcted myocardium. The results of this investigation revealed that the combined strategy of cell suspension injection and implantation of the cardiac patch on the infarct resulted in greater LV ED volume and scar thickness compared to cell suspension injection alone, at a 10 ± 3.5 months follow-up. Both treatments improved ejection fraction in those patients compared with the baseline measurements performed 1 week before coronary artery bypass (Chachques et al. 2007; Chachques et al. 2008).Clearly, the implantation of biomaterial scaffolds onto the infarct zone significantly contributed to the maintenance of LV wall thickness, probably by increasing the retention of transplanted cells as well as by providing space and support for penetrating host cells. Together, this resulted in thicker scars and reduced stress on LV wall, eventually attenuating remodeling of the heart and improving diastolic function.

One drawback of cardiac patches is the invasive procedure required for these approaches. In contrast, some injectable materials can be injected minimally invasively using catheter delivery. Several studies using alginate biomaterial have been initiated. Our encouraging results on *in situ* forming alginate hydrogel have led to a first-in-man clinical trial, proving the safety of intracoronary injection of alginate biomaterial in acute MI patients (BioLineRx Ltd. 2007) Due to the localized effect of the biomaterial, it received FDA approval for a regulatory path as a medical device, intended to be injected to patients following acute myocardial infarction, for prevention of ventricular remodeling and subsequent congestive heart failure. At present, this material is further examined in phase II placebo-controlled, multi-center, randomized double-blind trial (PRESERVATION I) (Ikaria Holdings Inc. 2010) Another injectable application of alginate, Algisyl-LVR™, is also under clinical investigation in patients with dilated cardiomyopathy (Lee et al. 2012). Here, the calcium cross-linked material is applied by 10–15 intramyocardial injections into the LV free wall in a circumferential pattern to thicken the LV wall and reshape the geometry of the LV, to improve LV function. Initial pilot, open label study was successfully completed, showing the safety and feasibility of the treatment, with signs of improved cardiac function (LoneStar Heart, Inc. 2009b) An Expanded randomized, controlled phase II study is currently underway (AUGMENT-HF) (LoneStar Heart, Inc. 2009a.).

Conclusions and Future Aspects

Biomaterials are likely to play an increasingly important role in the treatment protocols of myocardial infarction and heart failure. In re-constructing the cardiac patch by cardiac tissue engineering, biomaterials provide critical support for the seeded cells and guide tissue organization and maturation. These cardiac patches, if appropriately pre-vascularized prior to implantation and with size matching LV dimensions, have a potential to replace the infarct, thus aiding cardiac repair. Future studies in this direction would probably focus on the design of biomimetic materials with elastic properties, degradability and biological specificity matching those of the heart matrix. In addition, implementation of perfusion bioreactors and vascularization strategies are required to effectively engineer the cardiac patch.

In parallel, the use of injectable a-cellular biomaterials as ECM replacement implants is gaining considerable attention due to the relative simplicity of this approach compared to patch implantation. Injectable a-cellular biomaterials, such as alginate, have shown to thicken the infarct, prevent LV remodeling and improve cardiac function after acute MI. In addition, the delivery of alginate via the intracoronary artery to the infarct has been shown to be safe in clinical studies. These encouraging results indicate that the use of a-cellular biomaterials as a treatment strategy for cardiac repair is a viable approach and should be further explored. One issue in need for a further investigation is whether the biomaterial's effect on functional improvement is long lasting.

Biomaterials, when appropriately designed, can provide a depot for the sustained delivery of therapeutic proteins to the infarcted heart. Multiple regenerating factors can be delivered in a sequential order to activate the different reparative processes in the infarcted heart. Moreover, these biomaterials with combination of pro-survival factors can protect transplanted cells from the aggressive environment and induce differentiation and integration with the host tissue. Such possibility is particularly exciting as recent clinical studies with cell transplantation have shown that most injected cells do not survive after transplantation and their main effect is possibly induced by the paracrine factors that they secrete (Maltais et al. 2010; Mirotsou et al. 2011).

The applicability of biomaterials in tissue engineering and regeneration is gaining credibility due to the enormous creativity of many researchers. We expect this field of research and development to increase and to provide well-performing bio-inspired materials with greater biocompatibility and adaptability to the dynamic microenvironment during healing, for better support and cell retention enhancement after transplantation, with more efficient delivery options by non-invasive methods.

Acknowledgements The research has been supported by a European Union grant (FP7 HEALTH-F2-2009-222995 INELPY), 3 grants from the Israel Science Foundation (52/99-1, 793/04 & 1368/08) and the Israel Ministry of Science, Culture and Sport. Prof. Cohen holds the Claire and Harold Oshry Professor Chair in Biotechnology.

References

Akhyari P, Kamiya H, Haverich A, Karck M, Lichtenberg A (2008) Myocardial tissue engineering: the extracellular matrix. Eur J Cardiothorac Surg 34(2):229–241. doi:S1010-7940(08)00362-X [pii]. (10.1016/j.ejcts.2008.03.062)

Barash Y, Dvir T, Tandeitnik P, Ruvinov E, Guterman H, Cohen S (2010) Electric field stimulation integrated into perfusion bioreactor for cardiac tissue engineering. Tissue Eng Part C Methods 16(6):1417–1426. doi:10.1089/ten.TEC.2010.0068

BioLineRx Ltd. (2007) Safety and Feasibility of the Injectable BL-1040 Implant. Study NCT00557531, 2007. http://www.ClinicalTrials.gov. Accessed 5 Dec 2012.

Brown MA, Iyer RK, Radisic M (2008) Pulsatile perfusion bioreactor for cardiac tissue engineering. Biotechnol Prog 24(4):907–920. doi:10.1002/btpr.11

Cardin AD, Weintraub HJ (1989) Molecular modeling of protein-glycosaminoglycan interactions. Arteriosclerosis 9(1):21–32

Carrier RL, Rupnick M, Langer R, Schoen FJ, Freed LE, Vunjak-Novakovic G (2002) Perfusion improves tissue architecture of engineered cardiac muscle. Tissue Eng 8(2):175–188. doi:10.1089/107632702753724950

Caspi O, Lesman A, Basevitch Y, Gepstein A, Arbel G, Habib IH, Gepstein L, Levenberg S (2007) Tissue engineering of vascularized cardiac muscle from human embryonic stem cells. Circ Res 100(2):263–272. doi:01.RES.0000257776.05673.ff [pii]. (10.1161/01.RES.0000257776.05673.ff)

Chachques JC, Trainini JC, Lago N, Masoli OH, Barisani JL, Cortes-Morichetti M, Schussler O, Carpentier A (2007) Myocardial Assistance by Grafting a New Bioartificial Upgraded Myocardium (MAGNUM clinical trial): one year follow-up. Cell Transplant 16(9):927–934

Chachques JC, Trainini JC, Lago N, Cortes-Morichetti M, Schussler O, Carpentier A (2008) Myocardial Assistance by Grafting a New Bioartificial Upgraded Myocardium (MAGNUM trial): clinical feasibility study. Ann Thorac Surg 85(3):901–908. doi:10.1016/j.athoracsur.2007.10.052

Chiu LL, Radisic M, Vunjak-Novakovic G (2010) Bioactive scaffolds for engineering vascularized cardiac tissues. Macromol Biosci 10(11):1286–1301. doi:10.1002/mabi.201000202

Chiu LL, Reis LA, Radisic M (2012) Controlled delivery of thymosin beta4 for tissue engineering and cardiac regenerative medicine. Ann N Y Acad Sci 1269(1):16–25. doi:10.1111/j.1749-6632.2012.06718.x

Christman KL, Lee RJ (2006) Biomaterials for the treatment of myocardial infarction. J Am Coll Cardiol 48(5):907–913

Christman KL, Fok HH, Sievers RE, Fang Q, Lee RJ (2004a) Fibrin glue alone and skeletal myoblasts in a fibrin scaffold preserve cardiac function after myocardial infarction. Tissue Eng 10(3-4):403–409

Christman KL, Vardanian AJ, Fang Q, Sievers RE, Fok HH, Lee RJ (2004b) Injectable fibrin scaffold improves cell transplant survival, reduces infarct expansion, and induces neovasculature formation in ischemic myocardium. J Am Coll Cardiol 44(3):654–660

Cohen S, Leor J (2004) Rebuilding broken hearts. Biologists and engineers working together in the fledgling field of tissue engineering are within reach of one of their greatest goals: constructing a living human heart patch. Sci Am 291(5):44–51

Conti E, Carrozza C, Capoluongo E, Volpe M, Crea F, Zuppi C, Andreotti F (2004) Insulin-like growth factor-1 as a vascular protective factor. Circulation 110:2260–2265

D'Amario D, Cabral-Da-Silva MC, Zheng H, Fiorini C, Goichberg P, Steadman E, Ferreira-Martins J, Sanada F, Piccoli M, Cappetta D, D'Alessandro DA, Michler RE, Hosoda T, Anastasia L, Rota M, Leri A, Anversa P, Kajstura J (2011) Insulin-like growth factor-1 receptor identifies a pool of human cardiac stem cells with superior therapeutic potential for myocardial regeneration. Circ Res 108(12):1467–1481. doi:10.1161/CIRCRESAHA.111.240648

Dahlmann J, Krause A, Moller L, Kensah G, Mowes M, Diekmann A, Martin U, Kirschning A, Gruh I, Drager G (2013) Fully defined in situ cross-linkable alginate and hyaluronic acid hydrogels for myocardial tissue engineering. Biomaterials 34(4):940–951. doi:10.1016/j.biomaterials.2012.10.008

Dai W, Wold LE, Dow JS, Kloner RA (2005) Thickening of the infarcted wall by collagen injection improves left ventricular function in rats: a novel approach to preserve cardiac function after myocardial infarction. J Am Coll Cardiol 46(4):714–719

Dar A, Shachar M, Leor J, Cohen S (2002) Optimization of cardiac cell seeding and distribution in 3D porous alginate scaffolds. Biotechnol Bioeng 80:305–312

Davis ME, Hsieh PC, Grodzinsky AJ, Lee RT (2005) Custom design of the cardiac microenvironment with biomaterials. Circ Res 97(1):8–15

Davis ME, Hsieh PC, Takahashi T, Song Q, Zhang S, Kamm RD, Grodzinsky AJ, Anversa P, Lee RT (2006) Local myocardial insulin-like growth factor 1 (IGF-1) delivery with biotinylated peptide nanofibers improves cell therapy for myocardial infarction. Proc Natl Acad Sci U S A 103(21):8155–8160. doi:0602877103 [pii]. (10.1073/pnas.0602877103)

Dobaczewski M, Gonzalez-Quesada C, Frangogiannis NG (2010) The extracellular matrix as a modulator of the inflammatory and reparative response following myocardial infarction. J Mol Cell Cardiol 48(3):504–511. doi:S0022-2828(09)00308-3 [pii]. (10.1016/j.yjmcc.2009.07.015)

Dvir T, Benishti N, Shachar M, Cohen S (2006) A novel perfusion bioreactor providing a homogenous milieu for tissue regeneration. Tissue Eng 12:2843–2852

Dvir T, Levy O, Shachar M, Granot Y, Cohen S (2007) Activation of the ERK1/2 cascade via pulsatile interstitial fluid flow promotes cardiac tissue assembly. Tissue Eng 13(9):2185–2193. doi:10.1089/ten.2006.0364

Dvir T, Kedem A, Ruvinov E, Levy O, Freeman I, Landa N, Holbova R, Feinberg MS, Dror S, Etzion Y, Leor J, Cohen S (2009) Prevascularization of cardiac patch on the omentum improves its therapeutic outcome. Proc Natl Acad Sci U S A 106(35):14990–14995. doi:0812242106 [pii]. (10.1073/pnas.0812242106)

Eitan Y, Sarig U, Dahan N, Machluf M (2009) A cellular cardiac extracellular matrix as a scaffold for tissue engineering: In-vitro cell support, remodeling and biocompatibility. Tissue Eng Part C Methods 16(4):671–683. doi:10.1089/ten.TEC.2009.0111

Eschenhagen T, Fink C, Remmers U, Scholz H, Wattchow J, Weil J, Zimmermann W, Dohmen HH, Schafer H, Bishopric N, Wakatsuki T, Elson EL (1997) Three-dimensional reconstitution of embryonic cardiomyocytes in a collagen matrix: a new heart muscle model system. FASEB J 11(8):683–694

Frantz S, Bauersachs J, Ertl G (2009) Post-infarct remodelling: contribution of wound healing and inflammation. Cardiovasc Res 81(3):474–481. doi:cvn292 [pii] 10.1093/cvr/cvn292

Freed LE, Guilak F, Guo XE, Gray ML, Tranquillo R, Holmes JW, Radisic M, Sefton MV, Kaplan D, Vunjak-Novakovic G (2006) Advanced tools for tissue engineering: scaffolds, bioreactors, and signaling. Tissue Eng 12(12):3285–3305. doi:10.1089/ten.2006.12.3285

Freeman I, Cohen S (2009) The influence of the sequential delivery of angiogenic factors from affinity-binding alginate scaffolds on vascularization. Biomaterials 30(11):2122–2131. doi:S0142-9612(08)01060-0 [pii]. 10.1016/j.biomaterials.2008.12.057

Freeman I, Kedem A, Cohen S (2008) The effect of sulfation of alginate hydrogels on the specific binding and controlled release of heparin-binding proteins. Biomaterials 29(22):3260–3268. doi:S0142-9612(08)00278-0 [pii]. 10.1016/j.biomaterials.2008.04.025

Furuta A, Miyoshi S, Itabashi Y, Shimizu T, Kira S, Hayakawa K, Nishiyama N, Tanimoto K, Hagiwara Y, Satoh T, Fukuda K, Okano T, Ogawa S (2006) Pulsatile cardiac tissue grafts using a novel three-dimensional cell sheet manipulation technique functionally integrates with the host heart, in vivo. Circ Res 98(5):705–712

Gao J, Liu J, Gao Y, Wang C, Zhao Y, Chen B, Xiao Z, Miao Q, Dai J (2011) A myocardial patch made of collagen membranes loaded with collagen-binding human vascular endothelial growth factor accelerates healing of the injured rabbit heart. Tissue Eng Part A. 17(21–22): 2739–2747. doi:10.1089/ten.TEA.2011.0105

Godier-Furnemont AF, Martens TP, Koeckert MS, Wan L, Parks J, Arai K, Zhang G, Hudson B, Homma S, Vunjak-Novakovic G (2011) Composite scaffold provides a cell delivery platform for cardiovascular repair. Proc Natl Acad Sci U S A 108(19):7974–7979. doi:1104619108 [pii]. (10.1073/pnas.1104619108)

Goumans M-J, de Boer TP, Smits AM, van Laake LW, van Vliet P, Metz CHG, Korfage TH, Kats KP, Hochstenbach R, Pasterkamp G, Verhaar MC, van der Heyden MAG, de Kleijn D, Mummery CL, van Veen TAB, Sluijter JPG, Doevendans PA (2008) TGF-[beta]1 induces efficient differentiation of human cardiomyocyte progenitor cells into functional cardiomyocytes in vitro. Stem Cell Res 1(2):138–149

Hao X, Silva EA, Mansson-Broberg A, Grinnemo KH, Siddiqui AJ, Dellgren G, Wardell E, Brodin LA, Mooney DJ, Sylven C (2007) Angiogenic effects of sequential release of VEGF-A(165) and PDGF-BB with alginate hydrogels after myocardial infarction. Cardiovasc Res 75(1):178–185

Haraguchi Y, Shimizu T, Yamato M, Okano T (2011) Regenerative therapies using cell sheet-based tissue engineering for cardiac disease. Cardiol Res Pract 2011:845170. doi:10.4061/2011/845170

Hausenloy DJ, Yellon DM (2009) Cardioprotective growth factors. Cardiovasc Res 83(2):179–194. doi:cvp062 [pii]. (10.1093/cvr/cvp062)

Hsieh PC, Davis ME, Gannon J, MacGillivray C, Lee RT (2006) Controlled delivery of PDGF-BB for myocardial protection using injectable self-assembling peptide nanofibers. J Clin Invest 116(1):237–248. doi:10.1172/JCI25878

Ifkovits JL, Tous E, Minakawa M, Morita M, Robb JD, Koomalsingh KJ, Gorman JH 3rd, Gorman RC, Burdick JA (2010) Injectable hydrogel properties influence infarct expansion and extent of postinfarction left ventricular remodeling in an ovine model. Proc Natl Acad Sci U S A 107(25):11507–11512. doi:1004097107 [pii]. (10.1073/pnas.1004097107)

Ikaria Holdings Inc. (2010) IK-5001 for the Prevention of Remodeling of the ventricle and congestive heart failure after acute myocardial infarction (PRESERVATION 1). Study NCT01226563, 2010. http://www.ClinicalTrials.gov. Accessed 5 Dec 2012

Iraqi W, Rossignol P, Angioi M, Fay R, Nuee J, Ketelslegers JM, Vincent J, Pitt B, Zannad F (2009) Extracellular cardiac matrix biomarkers in patients with acute myocardial infarction complicated by left ventricular dysfunction and heart failure: insights from the eplerenone post-acute myocardial infarction heart failure efficacy and survival study (EPHESUS) study. Circulation 119(18):2471–2479. doi:10.1161/circulationaha.108.809194

Jugdutt BI (2003) Ventricular remodeling after infarction and the extracellular collagen matrix: when is enough enough? Circulation 108(11):1395–1403. doi:10.1161/01.CIR.0000085658.98621.49 108/11/1395 [pii]

Kutschka I, Chen IY, Kofidis T, Arai T, von Degenfeld G, Sheikh AY, Hendry SL, Pearl J, Hoyt G, Sista R, Yang PC, Blau HM, Gambhir SS, Robbins RC (2006) Collagen matrices enhance survival of transplanted cardiomyoblasts and contribute to functional improvement of ischemic rat hearts. Circulation 114(1 Suppl):I167–I173. doi:114/1_suppl/I-167 [pii]. (10.1161/CIRCULATIONAHA.105.001297)

Landa N, Miller L, Feinberg MS, Holbova R, Shachar M, Freeman I, Cohen S, Leor J (2008) Effect of injectable alginate implant on cardiac remodeling and function after recent and old infarcts in rat. Circulation 117(11):1388–1396. doi:10.1161/circulationaha.107.727420

Lee H, Cusick RA, Utsunomiya H, Ma PX, Langer R, Vacanti JP (2003) Effect of implantation site on hepatocytes heterotopically transplanted on biodegradable polymer scaffolds. Tissue Eng 9(6):1227–1232. doi:10.1089/10763270360728134

Lee RJ, Hinson A, Helgerson S, Bauernschmitt R, Sabbah HN (2012) Polymer-based restoration of left ventricular mechanics. Cell Transplant. 22(3): 529–533. doi:10.3727/096368911X637461

Leor J, Aboulafia-Etzion S, Dar A, Shapiro L, Barbash IM, Battler A, Granot Y, Cohen S (2000) Bioengineered cardiac grafts. A new approach to repair the infarcted myocardium? Circulation 102(supplIII):56–61

Leor J, Tuvia S, Guetta V, Manczur F, Castel D, Willenz U, Petnehazy O, Landa N, Feinberg MS, Konen E, Goitein O, Tsur-Gang O, Shaul M, Klapper L, Cohen S (2009) Intracoronary injection of in situ forming alginate hydrogel reverses left ventricular remodeling after myocardial infarction in Swine. J Am Coll Cardiol 54 (11):1014–1023. doi:S0735-1097(09)02060-9 [pii]. (10.1016/j.jacc.2009.06.010)

Lesman A, Habib M, Caspi O, Gepstein A, Arbel G, Levenberg S, Gepstein L (2009) Transplantation of a tissue-engineered human vascularized cardiac muscle. Tissue Eng Part A. 16(1): 115–125. doi:10.1089/ten.TEA.2009.0130

Li Q, Li B, Wang X, Leri A, Jana KP, Liu Y, Kajstura J, Baserga R, Anversa P (1997) Overexpression of insulin-like growth factor-1 in mice protects from myocyte death after infarction, attenuating ventricular dilation, wall stress, and cardiac hypertrophy. J Clin Invest 100(8):1991–1999. doi:10.1172/JCI119730

Li RK, Jia ZQ, Weisel RD, Mickle DA, Choi A, Yau TM (1999) Survival and function of bioengineered cardiac grafts. Circulation 100(19 Suppl):II63–II69

LoneStar Heart, Inc. (2009a) A randomized, controlled study to Evaluate Algisyl-LVR™ as a method of left ventricular augmentation for heart failure (AUGMENT-HF). Study NCT01311791, 2011. http://www.ClinicalTrials.gov. Accessed 5 Dec 2012

LoneStar Heart, Inc. (2009b) Safety and feasibility of Algisyl-LVR™ as a method of left ventricular restoration in patients with DCM undergoing open-heart surgery. Study NCT00847964, 2009. http://www.ClinicalTrials.gov. Accessed 5 Dec 2012

Lu H, Xu X, Zhang M, Cao R, Brakenhielm E, Li C, Lin H, Yao G, Sun H, Qi L, Tang M, Dai H, Zhang Y, Su R, Bi Y, Cao Y (2007) Combinatorial protein therapy of angiogenic and arteriogenic factors remarkably improves collaterogenesis and cardiac function in pigs. Proc Natl Acad Sci U S A 104(29):12140–12145. doi:0704966104 [pii]. (10.1073/pnas.0704966104)

Madden LR, Mortisen DJ, Sussman EM, Dupras SK, Fugate JA, Cuy JL, Hauch KD, Laflamme MA, Murry CE, Ratner BD (2010) Proangiogenic scaffolds as functional templates for cardiac tissue engineering. Proc Natl Acad Sci U S A 107(34):15211–15216. doi:1006442107 [pii]. (10.1073/pnas.1006442107)

Maidhof R, Marsano A, Lee EJ, Vunjak-Novakovic G (2010) Perfusion seeding of channeled elastomeric scaffolds with myocytes and endothelial cells for cardiac tissue engineering. Biotechnol Prog 26(2):565–572. doi:10.1002/btpr.337

Maltais S, Tremblay JP, Perrault LP, Ly HQ (2010) The paracrine effect: pivotal mechanism in cell-based cardiac repair. J Cardiovasc Transl Res 3(6):652–662. doi:10.1007/s12265-010-9198-2

Masuda S, Shimizu T, Yamato M, Okano T (2008) Cell sheet engineering for heart tissue repair. Adv Drug Deliv Rev 60(2):277–285. doi:S0169-409X(07)00249-9 [pii]. (10.1016/j.addr.2007.08.031)

Mirotsou M, Jayawardena TM, Schmeckpeper J, Gnecchi M, Dzau VJ (2011) Paracrine mechanisms of stem cell reparative and regenerative actions in the heart. J Mol Cell Cardiol 50(2):280–289

Miyagi Y, Chiu LL, Cimini M, Weisel RD, Radisic M, Li RK (2011) Biodegradable collagen patch with covalently immobilized VEGF for myocardial repair. Biomaterials 32(5):1280–1290. doi:S0142-9612(10)01295-0 [pii]. (10.1016/j.biomaterials.2010.10.007)

Mukherjee R, Zavadzkas JA, Saunders SM, McLean JE, Jeffords LB, Beck C, Stroud RE, Leone AM, Koval CN, Rivers WT, Basu S, Sheehy A, Michal G, Spinale FG (2008) Targeted myocardial microinjections of a biocomposite material reduces infarct expansion in pigs. Ann Thorac Surg 86(4):1268–1276. doi:S0003-4975(08)00955-7 [pii]. (10.1016/j.athoracsur.2008.04.107)

Nakamura T, Matsumoto K, Mizuno S, Sawa Y, Matsuda H (2005) Hepatocyte growth factor prevents tissue fibrosis, remodeling, and dysfunction in cardiomyopathic hamster hearts. Am J Physiol Heart Circ Physiol 288(5):H2131–H2139. doi:288/5/H2131 [pii]. (10.1152/ajpheart.01239.2003)

Nakamura T, Mizuno S, Matsumoto K, Sawa Y, Matsuda H (2000) Myocardial protection from ischemia/reperfusion injury by endogenous and exogenous HGF. J Clin Invest 106(12):1511–1519. doi:10.1172/JCI10226

Nelson WD, Zenovich AG, Ott HC, Stolen C, Caron GJ, Panoskaltsis-Mortari A, Barnes SA III, Xin X, Taylor DA (2007) Sex-Dependent Attenuation of Plaque Growth After Treatment With Bone Marrow Mononuclear Cells. Circ Res 101(12):1319–1327. doi:10.1161/circresaha.107.155564

Ott HC, Matthiesen TS, Goh SK, Black LD, Kren SM, Netoff TI, Taylor DA (2008) Perfusion-decellularized matrix: using nature's platform to engineer a bioartificial heart. Nat Med 14(2):213–221. doi:nm1684 [pii]. (10.1038/nm1684)

Perets A, Baruch Y, Weisbuch F, Shoshany G, Neufeld G, Cohen S (2003) Enhancing the vascularization of three-dimensional porous alginate scaffolds by incorporating controlled release basic fibroblast growth factor microspheres. J Biomed Mater Res A 65(4):489–497. doi:10.1002/jbm.a.10542

Radisic M, Park H, Shing H, Consi T, Schoen FJ, Langer R, Freed LE, Vunjak-Novakovic G (2004a) Functional assembly of engineered myocardium by electrical stimulation of cardiac myocytes cultured on scaffolds. Proc Natl Acad Sci U S A 101(52):18129–18134. doi:0407817101 [pii]. (10.1073/pnas.0407817101)

Radisic M, Yang L, Boublik J, Cohen RJ, Langer R, Freed LE, Vunjak-Novakovic G (2004b) Medium perfusion enables engineering of compact and contractile cardiac tissue. Am J Physiol Heart Circ Physiol 286(2):H507–516. doi:10.1152/ajpheart.00171.2003 00171.2003 [pii]

Radisic M, Park H, Gerecht S, Cannizzaro C, Langer R, Vunjak-Novakovic G (2007) Biomimetic approach to cardiac tissue engineering. Philos Trans R Soc Lond B Biol Sci 362(1484):1357–1368. doi:P083H57187563677 [pii]. (10.1098/rstb.2007.2121)

Radisic M, Marsano A, Maidhof R, Wang Y, Vunjak-Novakovic G (2008) Cardiac tissue engineering using perfusion bioreactor systems. Nat Protoc 3(4):719–738. doi:nprot.2008.40 [pii]. (10.1038/nprot.2008.40)

Richardson TP, Peters MC, Ennett AB, Mooney DJ (2001) Polymeric system for dual growth factor delivery. Nat Biotechnol 19(11):1029–1034. doi:10.1038/nbt1101-1029 nbt1101-1029 [pii]

Rosso F, Giordano A, Barbarisi M, Barbarisi A (2004) From cell-ECM interactions to tissue engineering. J Cell Physiol 199(2):174–180. doi:10.1002/jcp.10471

Ruvinov E, Dvir T, Leor J, Cohen S (2008) Myocardial repair: from salvage to tissue reconstruction. Expert Rev Cardiovasc Ther 6(5):669–686. doi:10.1586/14779072.6.5.669

Ruvinov E, Leor J, Cohen S (2010) The effects of controlled HGF delivery from an affinity-binding alginate biomaterial on angiogenesis and blood perfusion in a hindlimb ischemia model. Biomaterials 31(16):4573–4582. doi:S0142-9612(10)00255-3 [pii]. (10.1016/j.biomaterials.2010.02.026)

Ruvinov E, Harel-Adar T, Cohen S (2011a) Bioengineering the infarcted heart by applying bio-inspired materials. J Cardiovasc Transl Res 4(5):559–574. doi:10.1007/s12265-011-9288-9

Ruvinov E, Leor J, Cohen S (2011b) The promotion of myocardial repair by the sequential delivery of IGF-1 and HGF from an injectable alginate biomaterial in a model of acute myocardial infarction. Biomaterials 32(2):565–578. doi:10.1016/j.biomaterials.2010.08.097

Saik JE, Gould DJ, Watkins EM, Dickinson ME, West JL (2011) Covalently immobilized platelet-derived growth factor-BB promotes angiogenesis in biomimetic poly(ethylene glycol) hydrogels. Acta Biomater 7(1):133–143. doi:S1742-7061(10)00388-0 [pii]. (10.1016/j.actbio.2010.08.018)

Sapir Y, Kryukov O, Cohen S (2011) Integration of multiple cell-matrix interactions into alginate scaffolds for promoting cardiac tissue regeneration. Biomaterials 32(7):1838–1847. doi:S0142-9612(10)01430-4 [pii]. (10.1016/j.biomaterials.2010.11.008)

Schaaf S, Shibamiya A, Mewe M, Eder A, Stohr A, Hirt MN, Rau T, Zimmermann WH, Conradi L, Eschenhagen T, Hansen A (2011) Human engineered heart tissue as a versatile tool in basic research and preclinical toxicology. PLoS One 6(10):e26397. doi:10.1371/journal.pone.0026397 PONE-D-11-08333 [pii]

Segers VF, Lee RT (2007) Local delivery of proteins and the use of self-assembling peptides. Drug Discov Today 12(13–14):561–568

Segers VF, Tokunou T, Higgins LJ, MacGillivray C, Gannon J, Lee RT (2007) Local delivery of protease-resistant stromal cell derived factor-1 for stem cell recruitment after myocardial infarction. Circulation 116(15):1683–1692

Sekine H, Shimizu T, Kosaka S, Kobayashi E, Okano T (2006) Cardiomyocyte bridging between hearts and bioengineered myocardial tissues with mesenchymal transition of mesothelial cells. J Heart Lung Transplant 25(3):324–332. doi:S1053-2498(05)00744-8 [pii]. (10.1016/j.healun.2005.09.017)

Sekine H, Shimizu T, Dobashi I, Matsuura K, Hagiwara N, Takahashi M, Kobayashi E, Yamato M, Okano T (2011) Cardiac cell sheet transplantation improves damaged heart function via superior cell survival in comparison with dissociated cell injection. Tissue Eng Part A. 17(23–24): 2973–2980. doi:10.1089/ten.tea.2010.0659

Shachar M, Cohen S (2003) Cardiac tissue engineering, ex-vivo: Design principles in biomaterials and bioreactors. Heart Fail Rev 8:271–276

Shachar M, Tsur-Gang O, Dvir T, Leor J, Cohen S (2011) The effect of immobilized RGD peptide in alginate scaffolds on cardiac tissue engineering. Acta Biomater 7(1):152–162. doi:S1742-7061(10)00354-5 [pii]. (10.1016/j.actbio.2010.07.034)

Shapiro L, Cohen S (1997) Novel alginate sponges for cell culture and transplantation. Biomaterials 18(8):583–590. doi:S0142961296001810 [pii]

Shimizu T, Yamato M, Isoi Y, Akutsu T, Setomaru T, Abe K, Kikuchi A, Umezu M, Okano T (2002) Fabrication of pulsatile cardiac tissue grafts using a novel 3-dimensional cell sheet manipulation technique and temperature-responsive cell culture surfaces. Circ Res 90(3):e40

Shimizu T, Yamato M, Kikuchi A, Okano T (2003) Cell sheet engineering for myocardial tissue reconstruction. Biomaterials 24(13):2309–2316. doi:S0142961203001108 [pii]

Shin H, Jo S, Mikos AG (2003) Biomimetic materials for tissue engineering. Biomaterials 24(24):4353–4364. doi:S0142961203003399 [pii]

Shriver Z, Liu D, Sasisekharan R (2002) Emerging views of heparan sulfate glycosaminoglycan structure/activity relationships modulating dynamic biological functions. Trends Cardiovasc Med 12(2):71–77. doi:S1050173801001505 [pii]

Spinale FG (2007) Myocardial matrix remodeling and the matrix metalloproteinases: influence on cardiac form and function. Physiol Rev 87(4):1285–1342. doi:87/4/1285 [pii]. (10.1152/physrev.00012.2007)

Suleiman MS, Singh RJ, Stewart CE (2007) Apoptosis and the cardiac action of insulin-like growth factor I. Pharmacol Ther 114(3):278–294. doi:S0163-7258(07)00049-6 [pii]. (10.1016/j.pharmthera.2007.03.001)

Suzuki R, Hattori F, Itabashi Y, Yoshioka M, Yuasa S, Manabe-Kawaguchi H, Murata M, Makino S, Kokaji K, Yozu R, Fukuda K (2009) Omentopexy enhances graft function in myocardial cell sheet transplantation. Biochem Biophys Res Commun 387(2):353–359. doi:S0006-291X(09)01358-8 [pii]. (10.1016/j.bbrc.2009.07.024)

Tandon N, Cannizzaro C, Chao PH, Maidhof R, Marsano A, Au HT, Radisic M, Vunjak-Novakovic G (2009) Electrical stimulation systems for cardiac tissue engineering. Nat Protoc 4(2):155–173. doi:nprot.2008.183 [pii]. (10.1038/nprot.2008.183)

Tomita N, Morishita R, Taniyama Y, Koike H, Aoki M, Shimizu H, Matsumoto K, Nakamura T, Kaneda Y, Ogihara T (2003) Angiogenic property of hepatocyte growth factor is dependent on upregulation of essential transcription factor for angiogenesis, ets-1. Circulation 107(10):1411–1417

Tsur-Gang O, Ruvinov E, Landa N, Holbova R, Feinberg MS, Leor J, Cohen S (2009) The effects of peptide-based modification of alginate on left ventricular remodeling and function after myocardial infarction. Biomaterials 30(2):189–195. doi:S0142-9612(08)00687-X [pii]. (10.1016/j.biomaterials.2008.09.018)

Tulloch NL, Muskheli V, Razumova MV, Korte FS, Regnier M, Hauch KD, Pabon L, Reinecke H, Murry CE (2011) Growth of engineered human myocardium with mechanical loading and vascular coculture. Circ Res 109(1):47–59. doi:CIRCRESAHA.110.237206 [pii]. (10.1161/CIRCRESAHA.110.237206)

Vandervelde S, van Luyn MJ, Tio RA, Harmsen MC (2005) Signaling factors in stem cell-mediated repair of infarcted myocardium. J Mol Cell Cardiol 39(2):363–376

Wainwright JM, Czajka CA, Patel UB, Freytes DO, Tobita K, Gilbert TW, Badylak SF (2009) Preparation of cardiac extracellular matrix from an intact porcine heart. Tissue Eng Part C Methods 16(3):525–532. doi:10.1089/ten.TEC.2009.0392

Wang Y, Ahmad N, Wani MA, Ashraf M (2004) Hepatocyte growth factor prevents ventricular remodeling and dysfunction in mice via Akt pathway and angiogenesis. J Mol Cell Cardiol 37(5):1041–1052. doi:S0022-2828(04)00275-5 [pii]. (10.1016/j.yjmcc.2004.09.004)

Wang T, Wu DQ, Jiang XJ, Zhang XZ, Li XY, Zhang JF, Zheng ZB, Zhuo R, Jiang H, Huang C (2009) Novel thermosensitive hydrogel injection inhibits post-infarct ventricle remodelling. Eur J Heart Fail 11(1):14–19. doi:hfn009 [pii]. (10.1093/eurjhf/hfn009)

Webster KA (2007) Programmed death as a therapeutic target to reduce myocardial infarction. Trends Pharmacol Sci 28(9):492–499. doi:S0165-6147(07)00183-6 [pii]. (10.1016/j.tips.2007.07.004)

Ye KY, Black LD 3rd (2011) Strategies for tissue engineering cardiac constructs to affect functional repair following myocardial infarction. J Cardiovasc Transl Res 4(5): 575–591. doi:10.1007/s12265-011-9303-1

Yu J, Christman KL, Chin E, Sievers RE, Saeed M, Lee RJ (2009) Restoration of left ventricular geometry and improvement of left ventricular function in a rodent model of chronic ischemic cardiomyopathy. J Thorac Cardiovasc Surg 137(1):180–187. doi:S0022-5223(08)01459-1 [pii]. (10.1016/j.jtcvs.2008.08.036)

Zhang QX, Magovern CJ, Mack CA, Budenbender KT, Ko W, Rosengart TK (1997) Vascular endothelial growth factor is the major angiogenic factor in omentum: mechanism of the omentum-mediated angiogenesis. J Surg Res 67(2):147–154. doi:S0022-4804(96)94983-5 [pii]. (10.1006/jsre.1996.4983)

Zimmermann WH, Schneiderbanger K, Schubert P, Didie M, Munzel F, Heubach JF, Kostin S, Neuhuber WL, Eschenhagen T (2002) Tissue engineering of a differentiated cardiac muscle construct. Circ Res 90(2):223–230

Zimmermann WH, Didie M, Doker S, Melnychenko I, Naito H, Rogge C, Tiburcy M, Eschenhagen T (2006a) Heart muscle engineering: an update on cardiac muscle replacement therapy. Cardiovasc Res 71(3):419–429. doi:S0008-6363(06)00151-9 [pii]. (10.1016/j.cardiores.2006.03.023)

Zimmermann WH, Melnychenko I, Wasmeier G, Didie M, Naito H, Nixdorff U, Hess A, Budinsky L, Brune K, Michaelis B, Dhein S, Schwoerer A, Ehmke H, Eschenhagen T (2006b) Engineered heart tissue grafts improve systolic and diastolic function in infarcted rat hearts. Nat med 12(4):452–458. doi:10.1038/nm1394

Zmora S, Glicklis R, Cohen S (2002) Tailoring the pore architecture in 3-D alginate scaffolds by controlling the freezing regime during fabrication. Biomaterials 23(20):4087–4094

Chapter 7
Cell Therapy of Acute Myocardial Infarction and Ischemic Cardiomyopathy: From Experimental Findings to Clinical Trials

Stefan P. Janssens

Abstract Rapid progress in cardiovascular cell biology and in stem cell research offers new therapeutic opportunities for patients with acute myocardial infarction or advanced chronic ischemic disease. During the past decade, the recognition of the inherent regenerative potential of the human heart and the discovery of novel cues to modulate intrinsic repair capacity using progenitor cells have transformed innovative therapies for patients with acute or chronic ischemic disease. In the aftermath of pioneering observations with mononuclear bone marrow cell injections documenting improved myocardial function recovery in the infarcted murine heart, more than thousand patients have been enrolled in early phase randomized clinical cell transfer studies.

Introduction

Most of these exploratory studies have investigated whether autologous mixed bone marrow cells (BMC) can enhance global LV function recovery beyond state-of-the-art treatment and have shown mixed efficacy and reassuring medium-term safety. Experience with selected progenitor cell populations, including different mesenchymal stem cells and cardiac resident cells, or with allogeneic strategies, is growing and spurring a second wave of cell-based exploratory trials. Parallel studies *in vitro* and in small and large animals have steadily expanded our understanding of biological actions of bone marrow-derived and cardiac resident progenitor cells in repair processes of the injured heart. New scientific discoveries helped refocus therapeutic interventions, with a better understanding of patients in need and patient-related co-variables that affect response to therapy. While most traditional therapies aim to stall the advance of debilitating disease, regenerative medicine strives to reverse it. In AMI patients with increased risk of developing heart failure, the first, large scale international, multi-center clinical outcome study is under way and will investigate if autologous mononuclear BMC infusion during the subacute phase of MI can

S. P. Janssens (✉)
Department of Cardiovascular Diseases, Gasthuisberg University Hospital, KU Leuven, UZ Heresstraat 49, 3000 Leuven, Belgium
e-mail: stefan.janssens@med.kuleuven.be

J. Hescheler, E. Hofer (eds.), *Adult and Pluripotent Stem Cells*,
DOI 10.1007/978-94-017-8657-7_7,

improve survival over a 2 year observation. For chronic ischemic disease, it has been unclear how we can best harness scientific progress in successful treatment strategies. In the wake of the apparent failure of skeletal myoblasts, emerging allogeneic cell products and culture-expanded, reprogrammed or prespecified cells are being tested for their capacity to enhance myocardial repair and possibly activate cellular regeneration of the broken heart. Similar paradigms that have fostered the remarkable clinical translation with BMC in AMI treatment are being explored in chronic myocardial disease with selected cell populations. Finally, recognition that only a fraction of cells delivered with contemporary, device-based technologies engraft in the ischemic myocardium has spurred the development of increasingly effective enabling technologies aimed at enhancing retention and engraftment of cells in the heart. Continued research efforts into mechanisms of cell therapy and tissue engineering will lead to the discovery of truly regenerative cell products, capable of sustaining force-generation and nutrient perfusion or identify critical paracrine factors, enabling targeted protein-based therapies.

The Need for Adjunctive Cell Therapy in Acute and Chronic Ischemic Heart Disease

Ischemic heart disease remains the number 1 cause of death worldwide with an estimated incidence of 1 death every 39 s, and the incidence of the devastating chronic spectrum of the disease continues to grow, posing formidable health care challenges in the years to come. Optimized reperfusion strategies for acute myocardial infarction followed by evidence-based medical and device therapies cannot prevent the irreversible loss of functional myocardium. It is estimated that about 14 % of the LV mass is lost in a medium-size infarction which can correspond to almost 1 billion individual cardiac myocytes. Also, the decrease in early mortality after myocardial infarction over the past 20 years has resulted in an increasing population of infarct survivors and a fundamental shift of heart failure etiology (Velagaleti et al. 2008) away from valvular or hypertensive heart disease. In most patients, salvage of ischemic myocardium is incomplete and results in adverse left ventricular (LV) remodeling over time with subsequent progression to heart failure (Ganame et al. 2010). The single most important determinant of survival is the extent of postinfarction LV dysfunction (Postinfarction Research Group 1983; Skali et al. 2005). In the SAVE study, progressive LV enlargement has clearly been documented beyond the first year, long after histological infarct repair is thought to be completed and is associated with adverse outcome (St John Sutton et al. 1997). The Overture trial confirmed that LV remodeling and measurable changes in ventricular size using serial echocardiography are a powerful predictor of outcome in a heart failure population with baseline ejection fraction < 30 % (Solomon et al. 2005). In addition, long-term follow up in the VALIANT-trial in more than 14,000 patients with heart failure and post-MI ejection fraction < 35 % as determined by echocardiography or contrast angiography, emphasized poor clinical outcome with rates of reinfarction, death, or readmission for heart failure approximating 30 % at 2 years despite contemporary background therapy (Pfeffer et al. 2003).

In contrast to these earlier echocardiographic studies, the availability of cardiac MRI technology allows for a more comprehensive, noninvasive assessment of LV dimensions, function, perfusion and infarct size within a single imaging modality in a single session. MRI measurements can easily be repeated with time in a standardized fashion allowing a very detailed phenotypic description of the patient's evolution. Serial MRI analysis is thus uniquely positioned to identify at the time of first presentation specific signatures of patients who will likely develop heart failure over time and to evaluate the added value of adjuvant cell-based therapies. In a cohort of 260 reperfused STEMI patients, we reported that baseline MI size but not its location per se was an independent predictor of adverse LV remodeling (Masci et al.) and that MI size >23 % of LV mass was associated with progressive LV dilatation, chamber distortion, and deteriorating LV function. Of interest, autopsy data suggest that the loss of one third of LV contractile mass increases the likelihood of developing severe heart failure and corresponds to a net reduction of almost 2 billion cardiac myocytes (Beltrami et al. 1994). Because contemporary evidence-based therapy is not primarily directed to revert the massive loss of contractile elements and vascular structures in the infarct region, regenerative medicine has emerged as an appealing new strategy to reduce the burden of cardiovascular ischemic disease.

Cell-Based Cardiac Repair After Acute Myocardial Infarction: A Paradigm for Translating Innovation into Clinical Application

Since the first report that exogenously administered bone marrow mononuclear progenitors in the infarcted murine myocardium are capable of significant functional repair in the aftermath of myocardial infarction (Orlic et al. 2001), clinical translation has witnessed a remarkable pace. Initial experimental observations in mice mainly focused on plasticity of exogenously injected BMCs and suggested that transdifferentiation beyond original lineage commitment and autocrine myocardial regeneration represent major mechanisms of benefit (Wagers and Weissman 2004; Orlic et al. 2001). However, direct cardiomyocyte transdifferentiation of exogenously administered bone marrow-derived mononuclear cells has been questioned in subsequent murine studies (Balsam et al. 2004; Bel et al. 2003; Murry et al. 2004) and fusion of transplanted stem cells with endogenous cardiomyocytes has been proposed as an alternative explanation for the observed effects (Alvarez-Dolado et al. 2003; Nygren et al. 2004). In the following years, most experimental reports focused on paracrine trophic actions of mononuclear BMC and the release of neovascularization enhancing, anti-apoptotic and cytoprotective cytokines and growth factors (Fig. 7.1; for review see Gnecchi et al. 2008). Several carefully controlled experimental studies in small and large animal models have evidenced increased capillary density in the infarct border zone, greater absolute or stress-induced

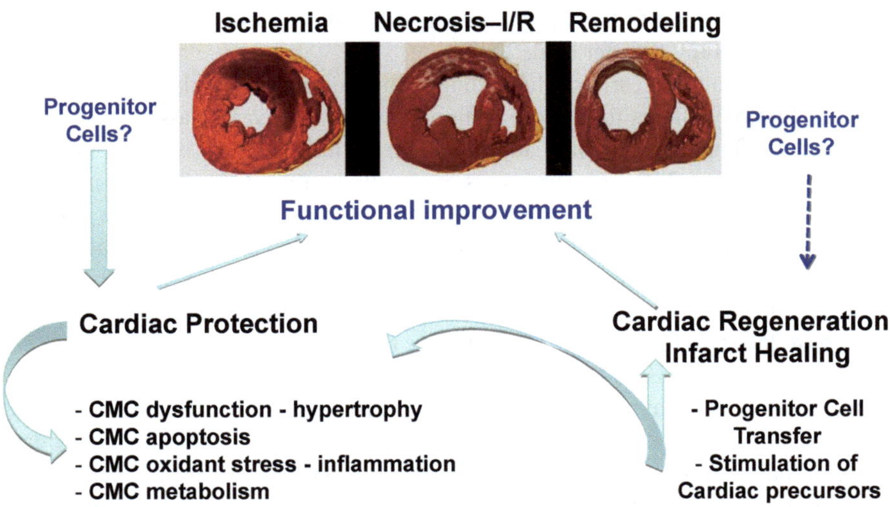

Fig. 7.1 Cell treatment in ischemic cardiomyopathy

myocardial nutrient blood flow and reduced apoptotic death rates of cardiac cells as putative mechanisms for the observed protection (Li et al. 2010; Moelker et al. 2006). Importantly, recent genetic fate mapping studies, using robust techniques to establish the derivatives of progenitor cell populations, suggest that c-kit-positive cells from the heart and cardiospheres do not differentiate into cardiac myocytes (Zaruba et al. 2010). Also c-kit derived cells form the bone marrow enhance cardiac myocyte renewal after myocardial infarction in mice without direct transdifferentiation (Loffredo et al. 2011; Andersen et al. 2009). Taken together, the data support important biological actions of paracrine factors released by the injected cells on a distinct population of endogenous cardiac progenitor cells.

Subsequent translation in post–AMI patients has been greatly facilitated by our extensive and clinically successful experience with BMC transplantation for hematological malignancies spanning several decades. In our own studies performed in the context of the FP7 collaborative "Infarct Cell Therapy" project, we have documented that patients receiving thrombolytic therapy for AMI have enhanced numbers of circulating CD34-positive hematopoietic progenitor cells during the early treatment period, suggesting a role for plasmin in mobilizing progenitor cells from bone marrow in response to remote ischemic injury (Tjwa et al. 2008). A similar transient increase in a number of distinct progenitor cell populations has been shown to parallel the clinical course of acute myocardial infarction, suggesting a potential reparative or reactive role for these cells in the context of severe tissue damage (Schomig et al. 2006; Wojakowski et al. 2009). Of note, some of the salutary mechanisms identified in experimental cardiovascular models have since been shown to operate in patients after intracoronary infusion of autologous BMCs (Erbs et al. 2007, 2006; Roncalli et al. 2011).

Bone Marrow Cell Therapy After Acute Myocardial Infarction

In an effort to translate the promising experimental results, most of the first wave of randomized, controlled clinical trials evaluating cell therapy in ischemic heart disease used autologous unfractionated mononuclear BMCs as a scalable, reproducible and easily accessible progenitor cell source. We will briefly review the most relevant findings in these studies and focus on how they helped zoom in on gaps in our knowledge and possible next steps in cardiac cell therapy.

The Surrogate Endpoint Discussion: From LV Dimensions and Ejection Fraction to Infarct Size and Myocardial Viability

The primary objectives in our quest for better treatment of patients with acute and chronic ischemic heart disease are to increase survival and improve quality of life (QOL). Because of the prohibitively large sample sizes required to demonstrate mortality benefit of a recently introduced cell-based therapy in patients with AMI, all initial major exploratory cell therapy trials relied on surrogate endpoints. The most critical and at times controversial aspect of such approach relates to the robustness of the chosen variable to truly reflect clinical efficacy. Long before the advent of cell therapy, many studies have reported that baseline LV dimensions and their derivative LVEF are significantly related to prognosis in patients with acute and chronic ischemic heart disease (Pfeffer and Pfeffer 1987; White et al. 1987; Cohn et al. 1993; Lee et al. 1993). Moreover, both the magnitude and direction of changes in LV dimensions and ejection fraction result in proportional changes in mortality (St John Sutton et al. 1997; Groenning et al. 2000), which is why it served arguably as the most frequently used substitute for the true clinical end point in the first wave of cell-based studies in AMI patients. The initial emphasis on load-dependent parameters including ejection fraction reflected a justified need for a workable, and affordable developmental surrogate in exploratory studies on safety, feasibility, risk/benefit ratio and in preliminary proof of efficacy trials. Following the introduction of comprehensive and highly accurate MR imaging, new surrogate endpoints including regional segmental function, myocardial viability and infarct size have been tested in large animal models with mixed results (Moelker et al. 2006; de Silva et al. 2008) and are progressively being introduced in clinical studies. Although it is still unclear to what extent changes in perfusion or viability reflect clinical outcome, accumulating data on myocardial infarct size suggest it may serve as a potentially more valuable surrogate endpoint, for which the direction and magnitude of change causes a proportional change in clinical outcome (Gibbons et al. 2004; Tjandrawidjaja et al. 2010). The progressively greater availability to integrate MR imaging in new exploratory trials using next generation cell products will enable to specifically focus on potential replacement of scar tissue by newly formed myocardial tissue.

However, solely relying on surrogates to assess a novel medicinal product, as complex as mixed mononuclear BMCs, carries the inherent risk of an incomplete or

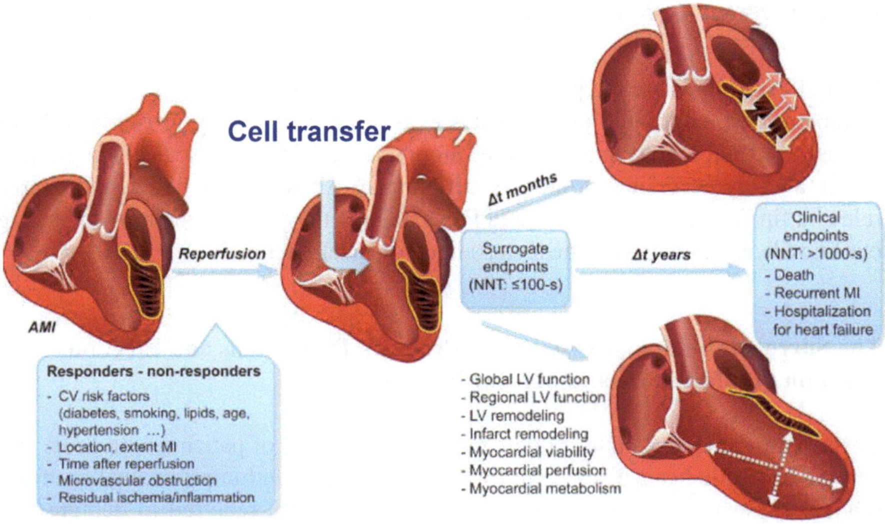

Fig. 7.2 Critical determinants for post-infarction cell-based therapy for myocardial repair

equivocal evaluation whereby such strategies could be either prematurely adopted or abandoned. These limitations account for some of the early skepticism on the significance of a relatively modest incremental increase in LVEF following cell therapy, and have jeopardized further development of a potentially promising treatment. Obviously, confirmation of the intrinsic added value of cell transfer can only come from adequately powered, multicenter clinical outcome studies (Fig. 7.2).

The First Randomized Controlled Trials Using BMCs After AMI

Approximately 5 years ago, four landmark trials (Table 7.1) reported the effect of unfractionated mononuclear BMCs as an adjunctive treatment strategy in patients suffering acute myocardial infarction. The major question was whether intracoronary infusion of mixed autologous BMCs in the aftermath of an acute myocardial infarction was capable of enhancing systolic function recovery beyond that obtained with state of the art interventional and pharmacological therapies.

The BOOST–trial randomly assigned 60 patients to intracoronary BMC infusion (n=30) or to standard therapy (control group, n=30) after successful percutaneous coronary intervention (PCI). Cell transfer, on average 4.8 days after PCI, was associated with a significant 6% increase in global LV ejection fraction (LVEF), assessed by MRI, 6 months later, and was attributable to improved regional contraction in the infarct border zone (Wollert et al. 2004). The reported incremental improvement in global LV systolic function in the overall study population at 6 months' follow-up was no longer statistically significant at 18 months and 5 years (Meyer et al. 2006, 2009). Post-hoc analyses suggested sustained BMC-mediated benefit in patients with larger infarcts and overall less diastolic dysfunction at 6 and

Table 7.1 Studies using BMCs as adjunctive treatment after myocardial infarction

Study	Design	Patient number	Baseline LVEF(imaging modality) (time after reperfusion)		Cell characteristics		Delivery Time	Primary endpoint
			Control	BMC	Number	Separation		BMC treatment effect
BOOST	– Prospective randomized	N=60	51±9%	50±10%	$24.6 \pm 9.4 \times 10^8$	4% Gelatin-Polysuccinate	4.8±1.3 d	Increase LVEF at 6 months
								6% increase in BMC *group over control group*
18 month follow-up	– Monocentre		(MRI)	(MRI)	–	–		Increase LVEF at 18 months No incremental increase over control group
5 year follow-up	– Single blind		(3.5±1.5 d)	(3.5±1.5 d)	–	–		Increase LVEF at 5 year No incremental increase over control group
Leuven-AMI	– Prospective randomized	N=67	47±8% (MRI) (3−5 d)	49±7% (MRI) (3−5 d)	$3.0 \pm 1.3 \times 10^8$	Ficoll	1 d	Increase LVEF at 4 months
1 year follow-up	– Monocentre				–	–		No incremental increase over control group Increase LVEF at 1 year
	– Double blind, placebo controlled							No incremental increase over control group
REPAIR-AMI	– Prospective randomized	N=204	47±10%	48±9%	$2.4 \pm 1.7 \times 10^8$	Ficoll-Hypaque	3−7 d	Increase LVEF at 4 months
1 year follow-up	– Multicentre (n =18)		(LV-Angio)	(LV-Angio)	–	–		*2.5% incremental increase in BMC group over control group*

Table 7.1 (continued)

Study	Design	Patient number	Baseline LVEF(imaging modality) (time after reperfusion)		Cell characteristics		Delivery Time	Primary endpoint
			Control	BMC	Number	Separation		BMC treatment effect
								Increase LVEF at 1 year
	– Double blind, placebo controlled		(3 − 7 d)	(3 − 7 d)				No significant increase in overall BMC group over control group, *6.6 % increase in subgroup with baseline LVEF < median*
ASTAMI	– Prospective randomized	*N* = 100	43 ± 12 % (SPECT) (4 − 7 d)	41 ± 0.4 % (SPECT) (4 − 7 d)	0.68 x 10^8 (median)	Ficoll-Lymphoprep	4 − 8 d	Increase LVEF at 6 months
	-Multicentre (*n* = 3)							
	-Single blind		54 ± 12 % (MRI) (2 − 3 w)	55 ± 14 % (MRI) (2 − 3 w)				
								No incremental increase over control group
1 year follow-up					–	–		Increase LVEF at 1 year
								No incremental increase over control group

18 months' follow up (Schaefer et al. 2006; Meyer et al. 2009). To control for the potentially confounding effect of procedure-related, repeated stop-flow conditions with transient ischemia and cytokine release on systolic LV function recovery, the Leuven AMI trial and the larger REPAIR-AMI trial studied the effect of adjunctive BMC infusion using a double-blind randomized design (Staat et al. 2005; Thibault et al. 2008).

In the Leuven AMI trial, we randomized 67 patients to intracoronary infusion of BMCs or placebo infusion within 24 h of successful PCI and used MRI to measure changes in LV function and structure from baseline to 4 months' follow-up. We did not observe an incremental recovery in global LV systolic function, the primary endpoint of the study (change in LVEF 2.2 % in control versus 3.4 % after BMC), although subgroup analysis suggested greater benefit in patients with large infarctions (Janssens et al. 2006). The latter observation is consistent with significant improvements in regional systolic function in the most severely infarcted segments, as evaluated by MRI and echocardiographic strain rate and with greater increases in metabolic activity patients with larger myocardial infarctions (Herbots et al. 2009). Importantly, infarct size (a prespecified secondary endpoint) was significantly lower in BMC-treated patients, suggesting a paracrine cytoprotective or immune-modulatory effect of early cell transfer after myocardial infarction. Of interest, a recent porcine study using magnetically labeled BMCs confirmed persistence and local distribution of the infused cells in the infarct border zone over a 6-week period and was associated with a significant infarct size reduction and global LV function enhancement (Graham et al. 2010).

In the REPAIR-AMI trial, 204 patients from 18 participating centers were randomly assigned to BMC or placebo infusion 3–7 days after successful PCI. LV angiography showed a incremental increase in global LVEF of 2.5 % at 4 months' follow-up in the BMC group (Schachinger et al. 2006a), an effect that was attributed to significantly better regional contraction in the infarct area and improved coronary Doppler flow signals (Erbs et al. 2007, 2006). Again, cell-mediated effects were more pronounced in segments with most severe contractile dysfunction, consistent with earlier observations, and no significant differences were documented in LV end-diastolic volume increase. In a subgroup of 54 patients, who were evaluated by serial MRI, a 2.8 % improvement in LVEF was reported at 1-year follow-up (Dill et al. 2009). One-year clinical follow-up data of the entire cohort showed significant reduction of the occurrence of major cardiovascular events in the BMC group (Schachinger et al. 2006b). These beneficial effects seemed sustained at extended follow-up but the relatively small number of patients mandates large-scale studies to confirm potential benefit of BMC transfer on hard clinical endpoints.

The "Autologous stem-cell transplantation in acute myocardial infarction" (ASTAMI)—trial randomized 100 patients with timely reperfused anterior MI's in three different centers to intracoronary BMC infusion 6 days later or to standard therapy. Global LV function recovery and infarct- and LV remodeling at 6 months and 1-year follow-up, as assessed by MRI, echocardiography or SPECT (single-photon emission computed tomography) were comparable irrespective of treatment assignment (Lunde et al. 2006, 2008). Functional impairment and lower numbers of infused BMCs (median 68×10^6 versus 198×10^6 in REPAIR-AMI and 172×10^6 in Leuven-AMI) or procedural differences in cell purification may contribute to the variable results in this study (Seeger et al. 2007).

In the FINnish stem CELL trial (FINCELL), the first study to evaluate BMC transfer in patients receiving thrombolysis, 80 patients were randomly assigned in two different centers (Huikuri et al. 2008) to intracoronary infusion of BMCs or placebo in a double blind trial design. Cell infusion, performed 2–6 days after thrombolysis at the time of PCI, resulted in an incremental improvement in global LVEF of ~6% as assessed by LV angiography at 6 months' follow-up. Paired echocardiography confirmed these results and no between group differences were observed in restenosis rate, arrhythmia risk, or clinical adverse events.

These first-generation randomized controlled clinical trials explored the safety and efficacy of intracoronary injection of autologous BMCs at variable time intervals following reperfusion. They mainly focused on global LVEF recovery as a surrogate marker for cell-mediated therapeutic response, and importantly, included patients with a moderate infarct size, unlikely to develop severe postinfarction heart failure. Overall, incremental global LV function recovery was modest, variable between trials and likely reflected significant procedure-related differences and variations in cell preparation and patient phenotype (Seeger et al. 2007). With a greater than 5 year clinical follow up period, the absence of significant cell-mediated in stent restenosis or thrombosis, arrhythmias, additional ischemic injury or intramyocardial neoplasms suggests a reassuring safety profile. These considerations together with a better appreciation of the unmet clinical need in patients with extensive myocardial damage, paved the way for additional RCT in patients with severely depressed LV function.

Second Generation Randomized Controlled Trials Using BMCs After AMI

The REGENT-trial was the first to prospectively select patients with greater ischemic burden and increased risk for developing heart failure compared to all previous trials (Tendera et al. 2009). The authors randomly assigned 200 patients with baseline LVEF <40% to a selected subpopulation of $CD34^+/CXCR4^+$ cells, to unselected BMC infusion or to standard therapy (control). The rational for selecting a subpopulation of $CD34^+$ $CXCR4^+$ cells stems from experimental observations in mice that interactions between the CXC chemokine, stromal cell-derived factor 1 (SDF-1), and its receptor CXC chemokine receptor 4 (CXCR4) on a subpopulation of bone marrow cells critically mediate ischemia-induced recruitment of bone marrow–derived progenitor cells to the myocardium (Penn 2010). $CXCR4^+$ cells are thus effectively mobilized from the bone marrow and home to the ischemic myocardium in response to the CXC chemokine stromal cell-derived factor 1 (SDF-1) (Tang et al. 2009), which is secreted by ischemic cardiac cells in the early post infarction period. Moreover, mobilized $CXCR4^+$ cells in patients with AMI have been shown to also express endothelial and cardiac lineage markers (Wojakowski et al. 2004), but transdifferentiation of these cells –if any- would be challenging to document in a clinical study. Serial MRI analysis in REGENT was available in 117 patients included in 5 different centers and paired analysis of either cell-treated group showed a 3% improvement in LVEF from baseline to 6-months' follow up. Preclinical studies have shown that SDF-1 chemokine levels return to baseline 7 days after the acute ischemic insult (Penn 2010), which corresponds to the median time interval of cell infusion in REGENT. In control patients, LVEF did not change over time despite contemporary background therapy, which is consistent with our earlier observations in patients with a large MI (Masci et al. 2011). When changes in LVEF over time were compared between the 3 groups, no significant differences in LVEF, LVEDV and LVESV were noted, possibly related to limited follow-up (the study was underpowered for its primary endpoint as <60% of patients completed the 6 months analysis) and imbalances in LVEF at baseline (Tendera et al. 2009). Adverse cardiovascular events were similar in all groups. As supported in previous studies, low baseline LVEF and longer time from onset of symptoms to PCI were predictors of cell-mediated improvement of LVEF.

No significant improvement in regional myocardial function, assessed by MRI at 4 months follow-up, was reported. Moreover, no significant differences were observed for secondary endpoints addressing global LVEF, LV remodeling and infarct remodeling. Of note, in contrast to the REPAIR-AMI-trial, no significant correlation was observed between cell-mediated LVEF improvement and the time from primary PCI to cell delivery. The failure to demonstrate improved regional function may relate to trial design, cell preparation protocol and uncertainties about progenitor cell functionality when stored in solutions containing concentrations of heparin that might affect cell migration, as shown in tumor cell invasion models (Bereczky et al. 2005).

The BONAMI-trial was the first to focus on metabolic effects of cell therapy and tested the capacity of BMCs to better protect myocardial viability after acute myocardial infarction (Roncalli et al.). This multicenter clinical trial in France randomized 101 patients with severe LV dysfunction (mean LVEF≈36%) to a control group or to intracoronary infusion of BMCs, at a mean interval of 9 days after primary PCI. SPECT-analysis at 3 months' follow-up showed a trend towards cell-mediated improvement in myocardial viability ($P=0.06$), the primary endpoint of this study. Interestingly, a multivariate analysis including major prognostic factors for LV function recovery after myocardial infarction suggested that smoking adversely affects cell transfer, while patients with microvascular obstruction derived a greater benefit from adjunctive cell therapy. The latter is consistent with earlier observations of enhanced BMC-mediated benefit in patients with large myocardial infarction, who are also more prone to develop microvascular obstruction.

Progenitor cell therapy for AMI in an autologous setting has a number of drawbacks. First, the need for cell harvesting, purification and—where needed—expansion implies significant costs, potentially limiting widespread use. Second, both bone marrow-derived and circulating progenitor cells from a diseased and elderly population may be functionally impaired and more prone to senesce (Heeschen et al. 2004; Vasa et al. 2001). In contrast, gradual recognition of the importance of paracrine mechanisms has favoured the concept of allogeneic cell transfer as a plausible alternative.

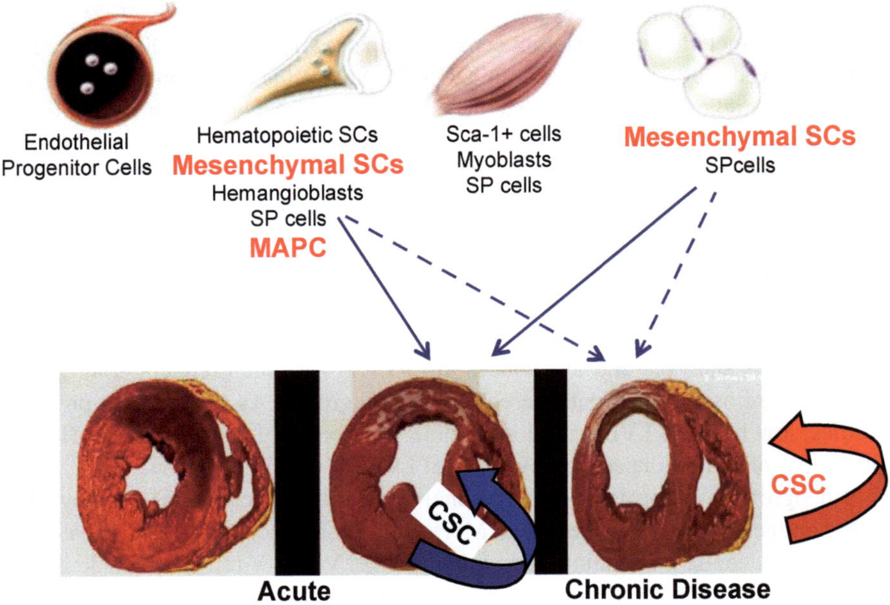

Fig. 7.3 Potential (Allogenic) cell sources for cardiac repair

Most of the attention has initially focused on mesenchymal or stromal cells, the first allogeneic cell type to be selected in clinical trials as an off-the-shelf readily available and scalable product (Hare et al. 2009). Mesenchymal stem cells are easily cultured progenitor cells that can be isolated from bone marrow, adipose tissue or skeletal muscle tissue and have multilineage potential. These cells can transdifferentiate into fat, cartilage, and bone in defined media (Pittenger et al. 1999; Crisan et al. 2009). MSCs are considered immune-privileged as they do not constitutively express immunogenic MHC class II or class Ia, but manifest immunosuppressive MHC class Ib and immune stimulatory properties upon stimulation by cytokines or in vivo injection (Poncelet et al. 2007; Nauta et al. 2006). In addition to their immunomodulatory properties, the cells secrete a complex mixture of growth factors, which in turn may enhance myocardial recovery in experimental myocardial infarction model by limiting apoptosis, inducing neovascularization and stimulating the endogenous pool of cardiac progenitor cells during the vulnerable peri- and post-infarction period (Mazo et al. 2012; Lee et al. 2009; Amado et al. 2005). To what extent MSC can enhance myocardial recovery after AMI via transdifferentiation into vascular and myogenic cells and whether or not MSCs can home to ischemic sites following systemic injection, remains controversial. Also, it is unknown if the purported distinct biological profile of adipose-tissue derived stem cells (ADSCs) or cells derived from the stroma vascular fraction would translate in greater clinical benefit (Mazo et al. 2011; Leobon et al. 2009; Fig. 7.3).

Selected Progenitor Cell Therapy After Acute Myocardial Infarction

Mesenchymal Stem Cells: Prototype of an Allogeneic Cell Transfer Approach

We should be careful extrapolating benefits observed with intramyocardial injections in mice to more clinically relevant intravascular delivery conditions. Preclinical studies in large animals have yielded conflicting results (Amado et al. 2005; Dubois et al. 2010; Schuleri et al. 2008, 2009; Mazo et al. 2012) and the prevailing view is that the degree of transdiffferentiation—if any—would not be sufficient to mediate observed beneficial effects. Moreover, the major limitation for intracoronary injection of these large and adhesive cells is their propensity to cause vascular obstruction (Ly et al. 2009) and dose-related additional myocardial injury. In pigs, transendocardial delivery of MSC were reported to stimulate the proliferation and differentiation of endogenous c-kit+ stem cells as part of their regenerative potential (Hatzistergos et al. 2010), although this effect was not reproduced in murine genetic lineage mapping studies (Loffredo et al. 2011). Because of potential risks of direct intramyocardial cell injection in patients in the acute phase of myocardial infarction, allogeneic MSCs were administered via slow IV infusion in patients 1–10 days after successful reperfusion of the infarcted territory (Hare et al. 2009). In this small study, the provisional safety endpoint was met, but larger trials are required for efficacy evaluation and long term follow up. Again, the finding of improved pulmonary function tests in patients randomized to MSC infusion is consistent with pulmonary capillary entrapment of cells, as observed in porcine imaging studies using 111In oxine labeled MSCs (Chin et al. 2003), with subsequent release of cytokines that modulate bronchial vasomotor tone. Alternatively, the Apollo trial was a double blind randomized placebo-controlled trial exploring the safety and feasibility of freshly isolated ADSC from liposuction aspirate using the Cellution™ system (Cytori Therapeutic Inc.) in patients with AMI (Meliga et al. 2007). To circumvent limitations of vascular delivery with MSCs or ADSCs, one study is currently investigating the safety and feasibility of transendocardial injection of 3 different doses (25, 75 and 150 M) allogeneic Stro-1+ mesenchymal precursor cells using the Nogastar™ mapping catheter and the Myostar™ injection catheter in patients with recent AMI (NCT00555828). Preliminary data from this single-blinded, dose-escalation, randomized, multicenter clinical trial were reported at the American College of Cardiology meeting in 2012 and confirmed feasibility, a reassuring safety profile and a favorable efficacy profile (E. Perin et al, ACC 2012 scientific sessions). Obviously, larger trials are required to better evaluate clinical efficacy of this allogeneic approach.

Multipotent Adult Progenitor Cells in AMI Patients

The identification and characterization of multipotent adult progenitor cells (MAPCs) with multipotent differentiation potential and a distinct gene-expression profile than MSC in the bone marrow has offered additional prospects for cardiac cell therapy (Jiang et al. 2002; Ulloa-Montoya et al. 2007). Encouraging safety and efficacy data in mice and pigs (Pelacho et al. 2007; Dimomeletis et al. 2010; Zeng et al. 2006, 2007; Van't Hof et al. 2007) led to the development of a proprietary allogeneic cell therapy product (MultiStem) and the first open label, multi-center dose escalation trial evaluating the safety of a single administration of Multistem cells following an AMI. In 19 patients Multistem was found to be safe following intracoronary adventitial injection in the aftermath of an AMI without development of prohibitive cellular or humoral immune responses. The treatment was well tolerated at all doses (20, 50 and 100 M cells) while indices of global LV function showed encouraging trends compared to baseline values (Penn et al. 2012). In view of the small number of patients, however, randomized, controlled Phase II studies need to be initiated to evaluate Multistem's ability to improve cardiac function after AMI.

Cardiac Resident Stem Cells: A Hidden Reserve Potential or Repair Tool?

Genuine regeneration of the myocardium occurs to varying degrees in a few animal species and results from robust proliferation of cardiac myocytes and possibly recruitment of epicardial progenitor cells, as clearly evidenced in zebrafish (Poss et al. 2002; Wills et al. 2008). Recently, the recognition of the adult mammalian heart as a dynamic organ, capable of cellular turnover of cardiac myocytes (with annual renewal rates ranging from between 0.5 and 1% to 10 and 40% depending on the organ's age and methodology used (Bergmann et al. 2009; Kajstura et al. 2010)) and limited regeneration through activation of endogenous cardiac progenitor cells and perhaps even recruitment of circulating progenitor cells has offered new perspectives for cardiac cell therapy (Anversa et al. 2006; Hosoda et al. 2010; Beltrami et al. 2003). In the post-natal injured mammalian heart, the focus now is on how to promote induction of cell cycle of already committed cardiac progenitor cells as a new strategy to prevent heart failure of the ischemic myocardium, rather than on an embryonic gene program of cell differentiation.

Much of the initial enthusiasm for bone marrow-derived progenitor cell therapy—to a large extent based on the presumed autocrine transdifferentiation paradigm (Orlic et al. 2001)—has been reinvigorated by the recent observation that exogenous BMC injection can activate endogenous progenitor cells with cardiogenic commitment (Loffredo et al. 2011). Using powerful genetic lineage mapping in transgenic mice, Loffredo and colleagues have reported that intramyocardial injection of bone marrow-derived c-kit+ cells—and not mesenchymal stem cells—can activate endogenous cardiac progenitors and improve LV function 8 weeks after myocardial infarction in mice (Loffredo et al. 2011). Of note, direct cardiomyocyte transdifferentiation by these exogenously administered bone marrow progenitors did not appear to mediate the observed benefit, as most of the injected exogenous cells failed to survive beyond the first week, recapitulating clinical experience with radiolabeled autologous human progenitor cells (Hofmann et al. 2005). One of the major questions remains to what extent intracoronary infusion of mononuclear bone marrow cells in patients after myocardial infarction could likewise boost endogenous cardiomyocyte precursor cell turnover and enhance functional recovery. Which progenitor cell types or released factors contribute to the effect, and the time for any measurable effect to appear in patients, still remain elusive and may account for the variable results of cell transfer in clinical studies.

Based on surface antigens and cardiogenic transcription markers, various subtypes of tissue-resident cardiac progenitor cells have been identified in the postnatal heart (Oh et al. 2003; Martin et al. 2008; Pfister et al. 2005; Moretti et al. 2006). Some of these markers are co-expressed, and represent a progression of cell stages with shared capacity to differentiate *in vitro* into the major constituent mature cell types of the heart, including endothelial cells, vascular smooth muscle cells and cardiac myocytes (Ellison et al. 2007; Nadal-Ginard and Fuster 2007). A clonal population of c-kit-expressing cells has been shown to occupy a distinct niche within the adult heart (Beltrami et al. 2003), but is notoriously sensitive to ischemic injury thereby compromising its endogenous repair capacity. Several laboratories have therefore invested time and effort in optimizing *in vitro* amplification protocols to obtain a sufficient number of cardiac progenitors from adult hearts after endomyocardial biopsy. Selected progenitor cells from digested biopsy specimens have been clonally expanded under permissive culture conditions (Ellison et al. 2011) or allowed to coalesce in culture to form 3-D spherical structures, called cardiospheres (Smith et al. 2007). Both clonally expanded c-kit+ cells and cardiospheres have a marked proliferative capacity, are capable of *in vitro* differentiation in contractile cardiac myocytes and have been shown to improve cardiac function in rodent and porcine models of AMI. Of interest, similar to earlier observations with mixed mononuclear BMCs, emerging data with cardiac-derived progenitors suggest that autocrine transdifferentiation is again a rare event and that the observed clinical benefit is mediated by paracrine trophic effects (Johnston et al. 2009; Andersen et al. 2009; Zaruba et al. 2010). An early-stage clinical phase 1 trial with intracoronary infusion of autologous culture-expanded cardiopsheres from a RV biopsy, was recently reported to be safe and capable of improving functional myocardial repair in patients with AMI (Makkar et al. 2012). Although still preliminary, the report of a reduction in scar size and a concomitant increase in viable myocardial tissue as a result of cell therapy offers an exciting prospect for regenerative strategies but need validation in larger prospective phase 2 trials. The only other clinical study using heart-derived progenitor cells (c-kit+ positive cells) is an interim analysis of an ongoing program of coronary bypass patients with persistent LV dysfunction despite surgical revascularization (Bolli et al. 2011). Proper evaluation of the effects of c-kit+ cell infusion will benefit from additional data including MRI analysis in the control group.

Endogenous Mobilization of Progenitor Cells

Bone marrow resident hematopoietic stem cells (HSC), with their inherent capacity for self-renewal and multilineage differentiation, are well known to support blood cell homeostasis by constitutively migrating out of the bone marrow and entering the blood. These cells express multiple traffic molecules and the chemokine receptor CXCR4, which facilitates their migration into non-lymphoid tissues where they survey extramedullary hematopoiesis (Wright et al. 2002). The paradigm that during tissue infection, this migratory pool of HSCs might act as a local source of highly proliferative progenitor cells capable of mounting an immediate innate immune response, has triggered trafficking-promoting strategies of BM-derived progenitor cells in an effort to thwart ischemic tissue injury. Therapeutic protocols using powerful mobilizing cytokines, including granulocyte-colony-stimulating factor (G-CSF), have been based on extensive experience of hematologists in stem cell transplantation and tested in several randomized controlled trials in AMI patients (Zohlnhofer et al. 2006; Ince et al. 2005; Valgimigli et al. 2005). Collectively, the data reported in early clinical trials totaling 445 AMI patients documented a reassuring safety profile but did not support an incremental functional benefit of G-CSF mobilization over background contemporary therapy (Zohlnhofer et al. 2008). The selection of patients and severity of the ischemic insult, dose and timing of growth factor administration, lack of understanding of local G-CSF receptor expression, choice of clinical or surrogate endpoints and investigative imaging modalities are generally considered to account for lack of incremental cardiac repair (Abdel-Latif et al. 2008; Zohlnhofer et al. 2008).

Interestingly, in a recent single blind, placebo-controlled trial, (Achilli et al. 2011) randomized 60 consecutive patients with anterior STEMI undergoing primary angioplasty within 12 h of symptom onset. Using magnetic resonance imaging for the primary and co-primary endpoints, they detected no change in LV ejection fraction from baseline (day 5) to 6 months' follow-up between groups, but a strikingly different pattern of LV remodeling and infarct healing with significant end-diastolic and end-systolic volume expansion in placebo-treated patients, versus virtually no change in G-CSF-treated patients. A critical comparison with prior studies suggests that perhaps the best window of opportunity for G-CSF mobilization strategies is the early post-revascularization phase in patients with large AMIs, who are at increased risk of developing maladaptive LV remodeling and heart failure. Although the small size of the study does not allow identifying underlying mechanisms, it is possible that the early timing of G-CSF administration is critically important, as suggested by the powerful cardioprotective effect of the cytokine via direct activation of the Jak-Stat pathway in cardiac myocytes (Harada et al. 2005). Whether these cell-specific, anti-apoptotic effects in the immediate aftermath of reperfusion predominate over any subsequent progenitor cell mobilization remains unknown at the present time and requires specifically designed larger clinical studies.

Finally, in an effort to increase stem cell homing following G-CSF administration, new pharmacological compounds modulating stem cell homing and mobilization cascades are being tested, including inhibitors of the dipeptidylpeptidase IV (DPP-IV), which cleaves and inactivates one of the major progenitor cell attracting chemokines, SDF-1a, and is currently in early clinical testing in AMI patients in the SITAGRAMI trial (EudraCT Number: 2007-003941-34) (Theiss et al. 2010). Other compounds, including the small molecule CXCR4 antagonist, AMD3100, have also been shown to augment the mobilization of endothelial progenitor cells from the bone marrow, which contributed to enhanced functional recovery of the infarcted murine heart and improved survival following a single-dose AMD3100 injection (Jujo et al. 2010). Of note, continuous AMD3100 infusion worsened outcome by blocking EPC incorporation despite higher circulating EPC levels. Early phase clinical trials in AMI patients are now being initiated using this agonist, which has been in clinical use for many years as the HIV therapeutic Plerixafor with reassuring safety profile.

Cell-Based Cardiac Repair for Chronic Myocardial Ischemia: A Paradigm for Translating Innovation into Clinical Application

Skeletal Myoblasts Versus Mononuclear Bone Marrow Cells

In contrast to AMI patients, our experience with cell-based therapies for chronic myocardial infarction and ischemic heart failure patients is more limited. Yet this patient population will undoubtedly increase over the years and become a major target for innovative therapies. The first larger scale multicenter, double blind randomized trial (MAGIC, Myoblast Autologous Grafting in Ischemic Cardiomyopathy) tested the effect of transepicardial injections of culture-expanded autologous myoblasts or placebo medium during coronary artery bypass grafting in 97 patients with severe ischemic LV dysfunction (Menasche et al. 2003). The trial was based on the concept that resident satellite cells in skeletal muscle can contribute to new myocytes in response to injury (Taylor et al. 1998) and was initiated following encouraging preclinical data and non-randomized clinical studies (Pouly et al. 2004; Menasche et al. 2003; Hagege et al. 2006). At 6 months' follow up the study failed to document incremental LV function recovery although the high dose myoblasts experienced a significant reduction in LV volumes and a favorable remodeling response. One major limitation is that we lack long-term follow up studies in representative large animal models. Moreover, the remaining uncertainties about potential arrhythmogenic side effects and the conflicting results with percutaneous endoventricular myoblasts injections or transcoronary venous injection in small scale studies (Macia and Boyden 2009; Siminiak et al. 2005) together with the high procedural costs have slowed further clinical development but will ultimately need clarification.

Whereas the potential pro-arrhythmic phenotype of engrafted skeletal myoblasts was thought to result from lack of gap junction formation with neighbouring cardiac myocytes, such electrical isolation was believed not to compromise potential efficacy of BMCs (Silva et al. 2011) or other selected cell types (MSC, ADSC, EPC) (Sato et al. 2011). The initial nonrandomized studies with intracoronary infusion of BMCs in patients more than 5 months after MI suggested that cell transfer may reduce infarct size and improve LV function and cardiac metabolism (Strauer et al. 2005), an effect that was sustained at 5 year's follow up and associated with improved quality of life and survival (Strauer et al. 2010). Because of the non-randomized design of the study, these results need to be confirmed in adequately powered, prospective randomized studies. Thus far, a carefully performed cross-over study of patients receiving intracoronary infusion of BMC more than 3 months after AMI showed that in comparison to background therapy these patients experienced a greater increase in LV ejection fraction caused by enhanced regional contraction in the area of cell administration (Assmus et al. 2006). One of the important remaining questions is which cell type can mediate myocardial repair during chronic ischemia and whether a single administration of progenitor cells will be sufficient for a sustained effect on myocardial repair or whether repeated injections will be preferable. Again, the lack of a suitable large animal model emphasizes the need for carefully designed clinical trials with detailed phenotypic analysis of residual myocardial perfusion, regional active deformation and viability. Such comprehensive analysis of the myocardial substrate of ischemic cardiomyopathy patients using multi-modal imaging technology is indispensable to evaluate the respective efficacy of new advanced medicinal cell products, including prespecified cardiopoietic mesenchymal stem cells (Behfar et al. 2010), MAPCs, or ADSCs in phase II trials. Finally, the SCIPIO trial is testing whether reinfusion of autologous c-kit+ cardiac stem cells isolated from the left atrial appendix during CABG will enhance LV functional recovery in patients with persistent severe LV dysfunction after surgical revascularization (Bolli et al. 2011). Early results confirm feasibility and safety but efficacy analysis mandates identical phenotypic evaluation in the control and in the cell-treated group.

Cell-Based Cardiac Repair for Ischemic Heart Disease: Challenges for Future Clinical Development and Key Issues

Cell therapy for myocardial infarction and chronic myocardial ischemia has come a long way during the past 2 decades. Collectively, available clinical data and supportive preclinical evidence suggest a promising role for mixed autologous BMC in the early post-infarction setting when LV function remains significantly impaired and the patient's clinical outcome is compromised. Cell therapy is mainly to be considered for those patients who truly represent an unmet clinical need and who's life expectancy is markedly compromised. One of the major remaining questions is whether or not BMCs can offer sufficient cardioprotection in the post-AMI setting to improve long-term clinical outcome. This challenge will be addressed in the incipient European BAMI study funded from the European Union Seventh Framework Programme for Research and Innovation (EU FP7 # 278967). The effect of intracoronary BMC infusion on all-cause mortality will be examined in a definitive 3,000 patient AMI population with severe residual LV dysfunction involving 21 partners in 11 European countries. The trial will aim to standardize the existing heterogeneity in cell isolation and purification protocols, provide rigorous quality control testing for all centers and will be powered to test a significant

BMC-mediated reduction in mortality at 2-y follow up. The trial results are expected to be announced in five years and will demonstrate whether this treatment saves lives.

While this study is underway, several parallel therapeutic strategies need to be further explored. Because true myocardial regeneration in sufficient quantities and within a short time frame to effectively replace the lost contractile mass remains an elusive goal, our best bet in AMI patients is to focus on effective myocardial salvage. Major clinical pharmacology-based trials specifically designed to reduce myocardial reperfusion injury and further loss of cardiac myocytes have not offered a substantial benefit (Hausenloy et al. 2010) and have spurred interest in innovative cell- or protein-based approaches. Obviously, the availability of an effective, off-the-shelf, readily available and scalable allogeneic advanced medicinal cell product or protein cocktail would be a formidable asset in our therapeutic armamentarium. Such biological intervention would be possible at the time of reperfusion or short thereafter, when the patient is most vulnerable and at maximal risk for additional myocardial damage. Our preclinical efforts should therefore be directed towards optimizing such therapeutic interventions by defining the optimal allogeneic cell product, protein cocktail or combination of both.

In advanced ischemic disease on the other hand, the major challenge remains to identify the fundamental criteria that govern successful and true myocardial regeneration and the replacement of the infarcted area with contractile tissue and sufficient nutrient blood supply. The question whether bone marrow- or cardiac-derived progenitor cells are more powerful in boosting myocardial regeneration remains unanswered, as well as the potential role for induced pluripotent stem cells (Takahashi et al. 2007). The latter cells are beyond doubt the most promising autologous pluripotent cell population and resemble embryonic stem cells. They can be maintained in culture for months and induced to differentiate into derivatives of all 3 germ layers. A major theoretical advantage of iPS cells was thought to be the absence of cellular immune rejection after in vivo administration. Although preclinical models have convincingly demonstrated the capacity to generate functional and electrically mature cardiomyocytes, which can integrate in the infarcted myocardium and restore LV function (Nelson et al. 2009), the general concern of tumor formation associated with cellular reprogramming limits clinical translation in its present form.

Cell Therapy: Key Issues

- There is a persistent unmet need for more effective therapies to compensate for loss of myocardial contractile performance in patients with acute and chronic myocardial ischemia.
- In AMI patients, intracoronary bone marrow cell (BMC) therapy is safe and mediates a modest improvement in LV function recovery.
- The mechanisms of the observed benefit of BMC are multifactorial, including stimulation of neovascularization, reduction of inflammation and cardiomyocyte

apoptosis, attraction of circulating progenitor cells and activation of endogenous cardiac progenitors.

- Mixed results in randomized clinical trials in AMI patients reflect procedural differences in cell preparation, patient- and infarct-related variables and comorbidities affecting progenitor cell phenotypes.
- Allogeneic cell transfer using mesenchymal, adipose-derived or cardiac-derived stem cells is an appealing concept but comparative efficacy studies with autologous strategies are lacking.
- Cumulative cardiomyocyte cell loss leading to severe LV dysfunction cannot yet be regenerated by exogenously transferred BMCs or isolated and culture-expanded cardiac stem cells. True myocardial regeneration requires high level engraftment and stable functional integration of a large number of transferred cells within a residual myocardial syncytium.
- Reprogramming of autologous somatic cells into inducible pluripotent stem cells with sufficient proliferative and regenerative potential to regenerate the required number of differentiated cardiac myocytes remains a distant objective, because uncertainties of karyotypic changes, chromosomal instability and risk of oncogenic drift (teratoma formation).
- Tissue engineered bioactive scaffolds in combination with iPS-derived contractile myocytes and vascular cells or directly reprogrammed somatic cells may represent the most promising future target for patients with chronic ischemic heart failure

References

Abdel-Latif A, Bolli R, Zuba-Surma EK, Tleyjeh IM, Hornung CA, Dawn B (2008) Granulocyte colony-stimulating factor therapy for cardiac repair after acute myocardial infarction: a systematic review and meta-analysis of randomized controlled trials. Am Heart J 156(2):216–226 e219. doi:S0002-8703(08)00277-9 [pii] 10.1016/j.ahj.2008.03.024

Achilli F, Malafronte C, Lenatti L, Gentile F, Dadone V, Gibelli G, Maggiolini S, Squadroni L, Di Leo C, Burba I, Pesce M, Mircoli L, Capogrossi MC, Di Lelio A, Camisasca P, Morabito A, Colombo G, Pompilio G (2011) Granulocyte colony-stimulating factor attenuates left ventricular remodelling after acute anterior STEMI: results of the single-blind, randomized, placebo-controlled multicentre STem cEll Mobilization in Acute Myocardial Infarction (STEM-AMI) Trial. Eur J Heart Fail 12(10):1111–1121. doi:hfq150 [pii] 10.1093/eurjhf/hfq150

Alvarez-Dolado M, Pardal R, Garcia-Verdugo JM, Fike JR, Lee HO, Pfeffer K, Lois C, Morrison SJ, Alvarez-Buylla A (2003) Fusion of bone-marrow-derived cells with Purkinje neurons, cardiomyocytes and hepatocytes. Nature 425(6961):968–973. doi:10.1038/nature02069 nature02069 [pii]

Amado LC, Saliaris AP, Schuleri KH, St John M, Xie JS, Cattaneo S, Durand DJ, Fitton T, Kuang JQ, Stewart G, Lehrke S, Baumgartner WW, Martin BJ, Heldman AW, Hare JM (2005) Cardiac repair with intramyocardial injection of allogeneic mesenchymal stem cells after myocardial infarction. Proc Natl Acad Sci U S A 102(32):11474–11479. doi:0504388102 [pii] 10.1073/pnas.0504388102

Andersen DC, Andersen P, Schneider M, Jensen HB, Sheikh SP (2009) Murine "cardiospheres" are not a source of stem cells with cardiomyogenic potential. Stem Cells 27(7):1571–1581. doi:10.1002/stem.72

Anversa P, Kajstura J, Leri A, Bolli R (2006) Life and death of cardiac stem cells: a paradigm shift in cardiac biology. Circulation 113(11):1451–1463

Assmus B, Walter DH, Lehmann R, Honold J, Martin H, Dimmeler S, Zeiher AM, Schachinger V (2006) Intracoronary infusion of progenitor cells is not associated with aggravated restenosis development or atherosclerotic disease progression in patients with acute myocardial infarction. Eur Heart J 27(24):2989–2995. doi:ehl235 [pii] 10.1093/eurheartj/ehl235

Balsam LB, Wagers AJ, Christensen JL, Kofidis T, Weissman IL, Robbins RC (2004) Haematopoietic stem cells adopt mature haematopoietic fates in ischaemic myocardium. Nature 428 (6983):668–673. doi:10.1038/nature02460 nature02460 [pii]

Behfar A, Yamada S, Crespo-Diaz R, Nesbitt JJ, Rowe LA, Perez-Terzic C, Gaussin V, Homsy C, Bartunek J, Terzic A (2010) Guided cardiopoiesis enhances therapeutic benefit of bone marrow human mesenchymal stem cells in chronic myocardial infarction. J Am Coll Cardiol 56(9):721–734. doi:S0735-1097(10)02236-9 [pii] 10.1016/j.jacc.2010.03.066

Bel A, Messas E, Agbulut O, Richard P, Samuel JL, Bruneval P, Hagege AA, Menasche P (2003) Transplantation of autologous fresh bone marrow into infarcted myocardium: a word of caution. Circulation 108(Suppl 1):II247–252. doi:10.1161/01.cir.0000089040.11131.d4 108/10_suppl_1/II-247 [pii]

Beltrami CA, Finato N, Rocco M, Feruglio GA, Puricelli C, Cigola E, Quaini F, Sonnenblick EH, Olivetti G, Anversa P (1994) Structural basis of end-stage failure in ischemic cardiomyopathy in humans. Circulation 89(1):151–163

Beltrami AP, Barlucchi L, Torella D, Baker M, Limana F, Chimenti S, Kasahara H, Rota M, Musso E, Urbanek K, Leri A, Kajstura J, Nadal-Ginard B, Anversa P (2003) Adult cardiac stem cells are multipotent and support myocardial regeneration. Cell 114(6):763–776. doi:S0092867403006871 [pii]

Bereczky B, Gilly R, Raso E, Vago A, Timar J, Tovari J (2005) Selective antimetastatic effect of heparins in preclinical human melanoma models is based on inhibition of migration and microvascular arrest. Clin Exp Metastasis 22(1):69–76. doi:10.1007/s10585-005-3859-6

Bergmann O, Bhardwaj RD, Bernard S, Zdunek S, Barnabe-Heider F, Walsh S, Zupicich J, Alkass K, Buchholz BA, Druid H, Jovinge S, Frisen J (2009) Evidence for cardiomyocyte renewal in humans. Science 324(5923):98–102. doi:324/5923/98 [pii] 10.1126/science.1164680

Bolli R, Chugh AR, D'Amario D, Loughran JH, Stoddard MF, Ikram S, Beache GM, Wagner SG, Leri A, Hosoda T, Sanada F, Elmore JB, Goichberg P, Cappetta D, Solankhi NK, Fahsah I, Rokosh DG, Slaughter MS, Kajstura J, Anversa P (2011) Cardiac stem cells in patients with ischaemic cardiomyopathy (SCIPIO): initial results of a randomised phase 1 trial. Lancet 378(9806):1847–1857. doi:10.1016/S0140-6736(11)61590-0

Chin BB, Nakamoto Y, Bulte JW, Pittenger MF, Wahl R, Kraitchman DL (2003) 111In oxine labelled mesenchymal stem cell SPECT after intravenous administration in myocardial infarction. Nucl Med Commun 24(11):1149–1154. doi:10.1097/01.mnm.0000101606.64255.03

Cohn JN, Johnson GR, Shabetai R, Loeb H, Tristani F, Rector T, Smith R, Fletcher R (1993) Ejection fraction, peak exercise oxygen consumption, cardiothoracic ratio, ventricular arrhythmias, and plasma norepinephrine as determinants of prognosis in heart failure. The V-HeFT VA cooperative studies group. Circulation 87(6 Suppl):VI5–16

Crisan M, Chen CW, Corselli M, Andriolo G, Lazzari L, Peault B (2009) Perivascular multipotent progenitor cells in human organs. Ann N Y Acad Sci 1176:118–123. doi:NYAS4967 [pii] 10.1111/j.1749-6632.2009.04967.x

de Silva R, Raval AN, Hadi M, Gildea KM, Bonifacino AC, Yu ZX, Yau YY, Leitman SF, Bacharach SL, Donahue RE, Read EJ, Lederman RJ (2008) Intracoronary infusion of autologous mononuclear cells from bone marrow or granulocyte colony-stimulating factor-mobilized apheresis product may not improve remodelling, contractile function, perfusion, or infarct size in a swine model of large myocardial infarction. Eur Heart J 29(14):1772–1782. doi:ehn216 [pii] 10.1093/eurheartj/ehn216

Dill T, Schachinger V, Rolf A (2009) Intracoronary administration of bone marrow-derived progenitor cells improves left ventricular function in patients at risk for adverse remodeling after acute ST-segment elevation myocardial infarction: Results of the Reinfusion of Enriched Progenitor cells And Infarct Remodeling in Acute Myocardial Infarction study (REPAIR-AMI) cardiac magnetic resonance imaging substudy. Am Heart J 157(3):541–547. doi: 10.1016/j.ahj.2008.11.011.http://wwwTCTMDcom

Dimomeletis I, Deindl E, Zaruba M, Groebner M, Zahler S, Laslo SM, David R, Kostin S, Deutsch MA, Assmann G, Mueller-Hoecker J, Feuring-Buske M, Franz WM (2010) Assessment of human MAPCs for stem cell transplantation and cardiac regeneration after myocardial infarction in SCID mice. Exp Hematol 38(11):1105–1114. doi:S0301-472X(10)00269-9 [pii] 10.1016/j.exphem.2010.06.013

Dubois C, Liu X, Claus P, Marsboom G, Pokreisz P, Vandenwijngaert S, Depelteau H, Streb W, Chaothawee L, Maes F, Gheysens O, Debyser Z, Gillijns H, Pellens M, Vandendriessche T, Chuah M, Collen D, Verbeken E, Belmans A, Van de Werf F, Bogaert J, Janssens S (2010) Differential effects of progenitor cell populations on left ventricular remodeling and myocardial neovascularization after myocardial infarction. J Am Coll Cardiol 55(20):2232–2243

Ellison GM, Torella D, Karakikes I, Nadal-Ginard B (2007) Myocyte death and renewal: modern concepts of cardiac cellular homeostasis. Nat Clin Pract Cardiovasc Med 4(Suppl 1):S52–59. doi:ncpcardio0773 [pii] 10.1038/ncpcardio0773

Ellison GM, Torella D, Dellegrottaglie S, Perez-Martinez C, Perez de Prado A, Vicinanza C, Purushothaman S, Galuppo V, Iaconetti C, Waring CD, Smith A, Torella M, Cuellas Ramon C, Gonzalo-Orden JM, Agosti V, Indolfi C, Galinanes M, Fernandez-Vazquez F, Nadal-Ginard B (2011) Endogenous cardiac stem cell activation by insulin-like growth factor-1/hepatocyte growth factor Intracoronary injection fosters survival and regeneration of the infarcted pig heart. J Am Coll Cardiol 58(9):977–986. doi:S0735-1097(11)01824-9 [pii] 10.1016/j.jacc.2011.05.013

Erbs S, Linke A, Schuler G, Hambrecht R (2006) Intracoronary administration of circulating blood-derived progenitor cells after recanalization of chronic coronary artery occlusion improves endothelial function. Circ Res 98(5):e48. doi:98/5/e48 [pii] 10.1161/01.RES.0000214407.58341.c8

Erbs S, Linke A, Schachinger V, Assmus B, Thiele H, Diederich KW, Hoffmann C, Dimmeler S, Tonn T, Hambrecht R, Zeiher AM, Schuler G (2007) Restoration of microvascular function in the infarct-related artery by intracoronary transplantation of bone marrow progenitor cells in patients with acute myocardial infarction: the Doppler Substudy of the Reinfusion of Enriched Progenitor Cells and Infarct Remodeling in Acute Myocardial Infarction (REPAIR-AMI) trial. Circulation 116(4):366–374. doi:CIRCULATIONAHA.106.671545 [pii] 10.1161/CIRCULATIONAHA.106.671545

Ganame J, Messalli G, Masci PG, Dymarkowski S, Abbasi K, Van de Werf F, Janssens S, Bogaert J (2010) Time course of infarct healing and left ventricular remodelling in patients with reperfused ST segment elevation myocardial infarction using comprehensive magnetic resonance imaging. Eur Radiol 21(4):693–701. doi:10.1007/s00330-010-1963-8

Gibbons RJ, Valeti US, Araoz PA, Jaffe AS (2004) The quantification of infarct size. J Am Coll Cardiol 44(8):1533–1542

Gnecchi M, Zhang Z, Ni A, Dzau VJ (2008) Paracrine mechanisms in adult stem cell signaling and therapy. Circ Res 103(11):1204–1219. doi:103/11/1204 [pii] 10.1161/CIRCRESAHA.108.176826

Graham JJ, Foltz WD, Vaags AK, Ward MR, Yang Y, Connelly KA, Vijayaraghavan R, Detsky JS, Hough MR, Stewart DJ, Wright GA, Dick AJ (2010) Long-term tracking of bone marrow progenitor cells following intracoronary injection post-myocardial infarction in swine using MRI. Am J Physiol Heart Circ Physiol 299(1):H125–133. doi:ajpheart.01260.2008 [pii] 10.1152/ajpheart.01260.2008

Groenning BA, Nilsson JC, Sondergaard L, Fritz-Hansen T, Larsson HB, Hildebrandt PR (2000) Antiremodeling effects on the left ventricle during beta-blockade with metoprolol in the treatment of chronic heart failure. J Am Coll Cardiol 36(7):2072–2080. doi:S0735-1097(00)01006-8 [pii]

Hagege AA, Marolleau JP, Vilquin JT, Alheritiere A, Peyrard S, Duboc D, Abergel E, Messas E, Mousseaux E, Schwartz K, Desnos M, Menasche P (2006) Skeletal myoblast transplantation in ischemic heart failure: long-term follow-up of the first phase I cohort of patients. Circulation 114(1 Suppl):I108–113. doi:114/1_suppl/I-108 [pii] 10.1161/CIRCULATIONAHA.105.000521

Harada M, Qin Y, Takano H, Minamino T, Zou Y, Toko H, Ohtsuka M, Matsuura K, Sano M, Nishi J, Iwanaga K, Akazawa H, Kunieda T, Zhu W, Hasegawa H, Kunisada K, Nagai T, Nakaya H, Yamauchi-Takihara K, Komuro I (2005) G-CSF prevents cardiac remodeling after myocardial infarction by activating the Jak-Stat pathway in cardiomyocytes. Nat Med 11(3):305–311

Hare JM, Traverse JH, Henry TD, Dib N, Strumpf RK, Schulman SP, Gerstenblith G, DeMaria AN, Denktas AE, Gammon RS, Hermiller JB, Jr., Reisman MA, Schaer GL, Sherman W (2009) A randomized, double-blind, placebo-controlled, dose-escalation study of intravenous adult human mesenchymal stem cells (prochymal) after acute myocardial infarction. J Am Coll Cardiol 54(24):2277–2286. doi:S0735-1097(09)03092-7 [pii] 10.1016/j.jacc.2009.06.055

Hatzistergos KE, Quevedo H, Oskouei BN, Hu Q, Feigenbaum GS, Margitich IS, Mazhari R, Boyle AJ, Zambrano JP, Rodriguez JE, Dulce R, Pattany PM, Valdes D, Revilla C, Heldman AW, McNiece I, Hare JM (2010) Bone marrow mesenchymal stem cells stimulate cardiac stem cell proliferation and differentiation. Circ Res 107(7):913–922. doi:CIRCRESAHA.110.222703 [pii] 10.1161/CIRCRESAHA.110.222703

Hausenloy DJ, Baxter G, Bell R, Botker HE, Davidson SM, Downey J, Heusch G, Kitakaze M, Lecour S, Mentzer R, Mocanu MM, Ovize M, Schulz R, Shannon R, Walker M, Walkinshaw G, Yellon DM (2010) Translating novel strategies for cardioprotection: the Hatter Workshop Recommendations. Basic Res Cardiol 105(6):677–686. doi:10.1007/s00395-010-0121-4

Heeschen C, Lehmann R, Honold J, Assmus B, Aicher A, Walter DH, Martin H, Zeiher AM, Dimmeler S (2004) Profoundly reduced neovascularization capacity of bone marrow mononuclear cells derived from patients with chronic ischemic heart disease. Circulation 109(13):1615–1622. doi:10.1161/01.CIR.0000124476.32871.E3 01.CIR.0000124476.32871.E3 [pii]

Herbots L, D'Hooge J, Eroglu E, Thijs D, Ganame J, Claus P, Dubois C, Theunissen K, Bogaert J, Dens J, Kalantzi M, Dymarkowski S, Bijnens B, Belmans A, Boogaerts M, Sutherland G, Van de Werf F, Rademakers F, Janssens S (2009) Improved regional function after autologous bone marrow-derived stem cell transfer in patients with acute myocardial infarction: a randomized, double-blind strain rate imaging study. Eur Heart J 30(6):662–670. doi:ehn532 [pii] 10.1093/eurheartj/ehn532

Hofmann M, Wollert KC, Meyer GP, Menke A, Arseniev L, Hertenstein B, Ganser A, Knapp WH, Drexler H (2005) Monitoring of bone marrow cell homing into the infarcted human myocardium. Circulation 111(17):2198–2202. doi:01.CIR.0000163546.27639.AA [pii] 10.1161/01.CIR.0000163546.27639.AA

Hosoda T, Kajstura J, Leri A, Anversa P (2010) Mechanisms of myocardial regeneration. Circ J 74(1):13–17. doi:JST.JSTAGE/circj/CJ-09-0665 [pii]

Huikuri HV, Kervinen K, Niemela M, Ylitalo K, Saily M, Koistinen P, Savolainen ER, Ukkonen H, Pietila M, Airaksinen JK, Knuuti J, Makikallio TH (2008) Effects of intracoronary injection of mononuclear bone marrow cells on left ventricular function, arrhythmia risk profile, and restenosis after thrombolytic therapy of acute myocardial infarction. Eur Heart J 29 (22):2723–2732. doi:ehn436 [pii] 10.1093/eurheartj/ehn436

Ince H, Petzsch M, Kleine HD, Eckard H, Rehders T, Burska D, Kische S, Freund M, Nienaber CA (2005) Prevention of left ventricular remodeling with granulocyte colony-stimulating factor after acute myocardial infarction: final 1-year results of the Front-Integrated Revascularization and Stem Cell Liberation in Evolving Acute Myocardial Infarction by Granulocyte Colony-Stimulating Factor (FIRSTLINE-AMI) Trial. Circulation 112(9 Suppl):I73–80. doi:112/9_suppl/I-73 [pii] 10.1161/CIRCULATIONAHA.104.524827

Janssens S, Dubois C, Bogaert J, Theunissen K, Deroose C, Desmet W, Kalantzi M, Herbots L, Sinnaeve P, Dens J, Maertens J, Rademakers F, Dymarkowski S, Gheysens O, Van Cleemput J, Bormans G, Nuyts J, Belmans A, Mortelmans L, Boogaerts M, Van de Werf F (2006) Autologous bone marrow-derived stem-cell transfer in patients with ST-segment elevation myocardial infarction: double-blind, randomised controlled trial. Lancet 367(9505):113–121. doi:S0140-6736(05)67861-0 [pii] 10.1016/S0140-6736(05)67861-0

Jiang Y, Jahagirdar BN, Reinhardt RL, Schwartz RE, Keene CD, Ortiz-Gonzalez XR, Reyes M, Lenvik T, Lund T, Blackstad M, Du J, Aldrich S, Lisberg A, Low WC, Largaespada DA,

Verfaillie CM (2002) Pluripotency of mesenchymal stem cells derived from adult marrow. Nature 418(6893):41–49. doi:10.1038/nature00870 nature00870 [pii]

Johnston PV, Sasano T, Mills K, Evers R, Lee ST, Smith RR, Lardo AC, Lai S, Steenbergen C, Gerstenblith G, Lange R, Marban E (2009) Engraftment, differentiation, and functional benefits of autologous cardiosphere-derived cells in porcine ischemic cardiomyopathy. Circulation 120(12):1075–1083, 1077 p following 1083. doi:CIRCULATIONAHA.108.816058 [pii] 10.1161/CIRCULATIONAHA.108.816058

Jujo K, Hamada H, Iwakura A, Thorne T, Sekiguchi H, Clarke T, Ito A, Misener S, Tanaka T, Klyachko E, Kobayashi K, Tongers J, Roncalli J, Tsurumi Y, Hagiwara N, Losordo DW (2010) CXCR4 blockade augments bone marrow progenitor cell recruitment to the neovasculature and reduces mortality after myocardial infarction. Proc Natl Acad Sci U S A 107(24):11008–11013. doi:0914248107 [pii]10.1073/pnas.0914248107

Kajstura J, Gurusamy N, Ogorek B, Goichberg P, Clavo-Rondon C, Hosoda T, D'Amario D, Bardelli S, Beltrami AP, Cesselli D, Bussani R, del Monte F, Quaini F, Rota M, Beltrami CA, Buchholz BA, Leri A, Anversa P (2010) Myocyte turnover in the aging human heart. Circ Res 107(11):1374–1386. doi:CIRCRESAHA.110.231498 [pii]10.1161/CIRCRESAHA.110.231498

Lee TH, Hamilton MA, Stevenson LW, Moriguchi JD, Fonarow GC, Child JS, Laks H, Walden JA (1993) Impact of left ventricular cavity size on survival in advanced heart failure. Am J Cardiol 72(9):672–676

Lee RH, Pulin AA, Seo MJ, Kota DJ, Ylostalo J, Larson BL, Semprun-Prieto L, Delafontaine P, Prockop DJ (2009) Intravenous hMSCs improve myocardial infarction in mice because cells embolized in lung are activated to secrete the anti-inflammatory protein TSG-6. Cell Stem Cell 5(1):54–63. doi:S1934-5909(09)00212-4 [pii] 10.1016/j.stem.2009.05.003

Leobon B, Roncalli J, Joffre C, Mazo M, Boisson M, Barreau C, Calise D, Arnaud E, Andre M, Puceat M, Penicaud L, Prosper F, Planat-Benard V, Casteilla L (2009) Adipose-derived cardiomyogenic cells: in vitro expansion and functional improvement in a mouse model of myocardial infarction. Cardiovasc Res 83(4):757–767. doi:cvp167 [pii] 10.1093/cvr/cvp167

Li CJ, Gaol RL, Yang YJ, Hu FH, Yang WX, You SJ, Song LF, Ruan YM, Qiao SB, Chen JL, Li JJ (2010) Effect of intracoronary infusion of bone marrow mononuclear cells or peripheral endothelial progenitor cells on myocardial ischemia-reperfusion injury in mini-swine. Chin Med Sci J 25(3):176–181

Loffredo FS, Steinhauser ML, Gannon J, Lee RT (2011) Bone marrow-derived cell therapy stimulates endogenous cardiomyocyte progenitors and promotes cardiac repair. Cell Stem Cell 8(4):389–398. doi:S1934-5909(11)00053-1 [pii] 10.1016/j.stem.2011.02.002

Lunde K, Solheim S, Aakhus S, Arnesen H, Abdelnoor M, Egeland T, Endresen K, Ilebekk A, Mangschau A, Fjeld JG, Smith HJ, Taraldsrud E, Grogaard HK, Bjornerheim R, Brekke M, Muller C, Hopp E, Ragnarsson A, Brinchmann JE, Forfang K (2006) Intracoronary injection of mononuclear bone marrow cells in acute myocardial infarction. N Engl J Med 355 (12):1199–1209

Lunde K, Solheim S, Forfang K, Arnesen H, Brinch L, Bjornerheim R, Ragnarsson A, Egeland T, Endresen K, Ilebekk A, Mangschau A, Aakhus S (2008) Anterior myocardial infarction with acute percutaneous coronary intervention and intracoronary injection of autologous mononuclear bone marrow cells: safety, clinical outcome, and serial changes in left ventricular function during 12-months' follow-up. J Am Coll Cardiol 51(6):674–676. doi:S0735-1097(07)03674-1 [pii] 10.1016/j.jacc.2007.10.032

Ly HQ, Hoshino K, Pomerantseva I, Kawase Y, Yoneyama R, Takewa Y, Fortier A, Gibbs-Strauss SL, Vooght C, Frangioni JV, Hajjar RJ (2009) In vivo myocardial distribution of multipotent progenitor cells following intracoronary delivery in a swine model of myocardial infarction. Eur Heart J 30(23):2861–2868. doi:ehp322 [pii] 10.1093/eurheartj/ehp322

Macia E, Boyden PA (2009) Stem cell therapy is proarrhythmic. Circulation 119(13):1814–1823. doi:119/13/1814 [pii] 10.1161/CIRCULATIONAHA.108.779900

Makkar RR, Smith RR, Cheng K, Malliaras K, Thomson LE, Berman D, Czer LS, Marban L, Mendizabal A, Johnston PV, Russell SD, Schuleri KH, Lardo AC, Gerstenblith G, Marban E (2012)

Intracoronary cardiosphere-derived cells for heart regeneration after myocardial infarction (CADUCEUS): a prospective, randomised phase 1 trial. Lancet 379(9819):895–904. doi:10.1016/S0140-6736(12)60195-0

Martin CM, Ferdous A, Gallardo T, Humphries C, Sadek H, Caprioli A, Garcia JA, Szweda LI, Garry MG, Garry DJ (2008) Hypoxia-inducible factor-2alpha transactivates Abcg2 and promotes cytoprotection in cardiac side population cells. Circ Res 102(9):1075–1081. doi:CIRCRESAHA.107.161729 [pii] 10.1161/CIRCRESAHA.107.161729

Masci PG, Ganame J, Francone M, Desmet W, Lorenzoni V, Iacucci I, Barison A, Carbone I, Lombardi M, Agati L, Janssens S, Bogaert J (2011) Relationship between location and size of myocardial infarction and their reciprocal influences on post-infarction left ventricular remodelling. Eur Heart J 32(13):1640–1648. doi:ehr064 [pii] 10.1093/eurheartj/ehr064

Mazo M, Gavira JJ, Pelacho B, Prosper F (2011) Adipose-derived stem cells for myocardial infarction. J Cardiovasc Transl Res 4(2):145–153. doi:10.1007/s12265-010-9246-y

Mazo M, Hernandez S, Gavira JJ, Abizanda G, Arana M, Lopez-Martinez T, Moreno C, Merino J, Martino-Rodriguez A, Uixeira A, de Jalon JA, Pastrana J, Martinez-Caro D, Prosper F (2012) Treatment of reperfused ischemia with adipose-derived stem cells in a preclinical swine model of myocardial infarction. Cell Transplant 21(12):2723–2733. doi:10.3727/096368912X638847

Meliga E, Strem BM, Duckers HJ, Serruys PW (2007) Adipose-derived cells. Cell Transplant 16(9):963–970

Menasche P, Hagege AA, Vilquin JT, Desnos M, Abergel E, Pouzet B, Bel A, Sarateanu S, Scorsin M, Schwartz K, Bruneval P, Benbunan M, Marolleau JP, Duboc D (2003) Autologous skeletal myoblast transplantation for severe postinfarction left ventricular dysfunction. J Am Coll Cardiol 41(7):1078–1083

Meyer GP, Wollert KC, Drexler H (2006) Stem cell therapy: a new perspective in the treatment of patients with acute myocardial infarction. Eur J Med Res 11(10):439–446

Meyer GP, Wollert KC, Lotz J, Pirr J, Rager U, Lippolt P, Hahn A, Fichtner S, Schaefer A, Arseniev L, Ganser A, Drexler H (2009) Intracoronary bone marrow cell transfer after myocardial infarction: 5-year follow-up from the randomized-controlled BOOST trial. Eur Heart J 30(24):2978–2984. doi:ehp374 [pii] 10.1093/eurheartj/ehp374

Moelker AD, Baks T, van den Bos EJ, van Geuns RJ, de Feyter PJ, Duncker DJ, van der Giessen WJ (2006) Reduction in infarct size, but no functional improvement after bone marrow cell administration in a porcine model of reperfused myocardial infarction. Eur Heart J 27(24):3057–3064. doi:ehl401 [pii] 10.1093/eurheartj/ehl401

Moretti A, Caron L, Nakano A, Lam JT, Bernshausen A, Chen Y, Qyang Y, Bu L, Sasaki M, Martin-Puig S, Sun Y, Evans SM, Laugwitz KL, Chien KR (2006) Multipotent embryonic isl1 + progenitor cells lead to cardiac, smooth muscle, and endothelial cell diversification. Cell 127(6):1151–1165. doi:S0092-8674(06)01398-5 [pii] 10.1016/j.cell.2006.10.029

Murry CE, Soonpaa MH, Reinecke H, Nakajima H, Nakajima HO, Rubart M, Pasumarthi KB, Virag JI, Bartelmez SH, Poppa V, Bradford G, Dowell JD, Williams DA, Field LJ (2004) Haematopoietic stem cells do not transdifferentiate into cardiac myocytes in myocardial infarcts. Nature 428(6983):664–668. doi:10.1038/nature02446 nature02446 [pii]

Nadal-Ginard B, Fuster V (2007) Myocardial cell therapy at the crossroads. Nat Clin Pract Cardiovasc Med 4(1):1. doi:ncpcardio0743 [pii] 10.1038/ncpcardio0743

Nauta AJ, Westerhuis G, Kruisselbrink AB, Lurvink EG, Willemze R, Fibbe WE (2006) Donor-derived mesenchymal stem cells are immunogenic in an allogeneic host and stimulate donor graft rejection in a nonmyeloablative setting. Blood 108(6):2114–2120. doi:blood-2005-11-011650 [pii] 10.1182/blood-2005-11-011650

Nelson TJ, Martinez-Fernandez A, Yamada S, Perez-Terzic C, Ikeda Y, Terzic A (2009) Repair of acute myocardial infarction by human stemness factors induced pluripotent stem cells. Circulation 120(5):408–416. doi:CIRCULATIONAHA.109.865154 [pii] 10.1161/CIRCULATIONAHA.109.865154

Nygren JM, Jovinge S, Breitbach M, Sawen P, Roll W, Hescheler J, Taneera J, Fleischmann BK, Jacobsen SE (2004) Bone marrow-derived hematopoietic cells generate cardiomyocytes at a low frequency through cell fusion, but not transdifferentiation. Nat Med 10(5):494–501. doi:10.1038/nm1040 nm1040 [pii]

Oh H, Bradfute SB, Gallardo TD, Nakamura T, Gaussin V, Mishina Y, Pocius J, Michael LH, Behringer RR, Garry DJ, Entman ML, Schneider MD (2003) Cardiac progenitor cells from adult myocardium: homing, differentiation, and fusion after infarction. Proc Natl Acad Sci U S A 100(21):12313–12318. doi:10.1073/pnas.2132126100 2132126100 [pii]

Orlic D, Kajstura J, Chimenti S, Jakoniuk I, Anderson SM, Li B, Pickel J, McKay R, Nadal-Ginard B, Bodine DM, Leri A, Anversa P (2001) Bone marrow cells regenerate infarcted myocardium. Nature 410 (6829):701–705. doi:10.1038/35070587 35070587 [pii]

Pelacho B, Nakamura Y, Zhang J, Ross J, Heremans Y, Nelson-Holte M, Lemke B, Hagenbrock J, Jiang Y, Prosper F, Luttun A, Verfaillie CM (2007) Multipotent adult progenitor cell transplantation increases vascularity and improves left ventricular function after myocardial infarction. J Tissue Eng Regen Med 1(1):51–59. doi:10.1002/term.7

Penn MS (2010) SDF-1:CXCR4 axis is fundamental for tissue preservation and repair. Am J Pathol 177(5):2166–2168. doi:S0002-9440(10)60269-6 [pii] 10.2353/ajpath.2010.100803

Penn MS, Ellis S, Gandhi S, Greenbaum A, Hodes Z, Mendelsohn FO, Strasser D, Ting AE, Sherman W (2012) Adventitial delivery of an allogeneic bone marrow-derived adherent stem cell in acute myocardial infarction: phase I clinical study. Circ Res 110(2):304–311. doi:10.1161/CIRCRESAHA.111.253427

Pfeffer MA, McMurray JJ, Velazquez EJ, Rouleau JL, Kober L, Maggioni AP, Solomon SD, Swedberg K, Van de Werf F, White H, Leimberger JD, Henis M, Edwards S, Zelenkofske S, Sellers MA, Califf RM (2003) Valsartan, captopril, or both in myocardial infarction complicated by heart failure, left ventricular dysfunction, or both. N Engl J Med 349(20):1893–1906. doi:10.1056/NEJMoa032292 NEJMoa032292 [pii]

Pfeffer MA, Pfeffer JM (1987) Ventricular enlargement and reduced survival after myocardial infarction. Circulation 75(5 Pt 2):IV93–97

Pfister O, Mouquet F, Jain M, Summer R, Helmes M, Fine A, Colucci WS, Liao R (2005) CD31- but Not CD31+ cardiac side population cells exhibit functional cardiomyogenic differentiation. Circ Res 97(1):52–61. doi:01.RES.0000173297.53793.fa [pii] 10.1161/01.RES.0000173297.53793.fa

Pittenger MF, Mackay AM, Beck SC, Jaiswal RK, Douglas R, Mosca JD, Moorman MA, Simonetti DW, Craig S, Marshak DR (1999) Multilineage potential of adult human mesenchymal stem cells. Science 284(5411):143–147

Poncelet AJ, Vercruysse J, Saliez A, Gianello P (2007) Although pig allogeneic mesenchymal stem cells are not immunogenic in vitro, intracardiac injection elicits an immune response in vivo. Transplantation 83(6):783–790. doi:10.1097/01.tp.0000258649.23081.a3 00007890-200703270-00021 [pii]

Poss KD, Wilson LG, Keating MT (2002) Heart regeneration in zebrafish. Science 298(5601):2188–2190. doi:10.1126/science.1077857 298/5601/2188 [pii]

Postinfarction Research Group M (1983) Risk stratification and survival after myocardial infarction. N Engl J Med 309(6):331–336. doi:10.1056/NEJM198308113090602

Pouly J, Hagege AA, Vilquin JT, Bissery A, Rouche A, Bruneval P, Duboc D, Desnos M, Fiszman M, Fromes Y, Menasche P (2004) Does the functional efficacy of skeletal myoblast transplantation extend to nonischemic cardiomyopathy? Circulation 110(12):1626–1631. doi:10.1161/01.CIR.0000142861.55862.15 01.CIR.0000142861.55862.15 [pii]

Roncalli J, Mouquet F, Piot C, Trochu JN, Le Corvoisier P, Neuder Y, Le Tourneau T, Agostini D, Gaxotte V, Sportouch C, Galinier M, Crochet D, Teiger E, Richard MJ, Polge AS, Beregi JP, Manrique A, Carrie D, Susen S, Klein B, Parini A, Lamirault G, Croisille P, Rouard H, Bourin P, Nguyen JM, Delasalle B, Vanzetto G, Van Belle E, Lemarchand P (2011) Intracoronary autologous mononucleated bone marrow cell infusion for acute myocardial infarction: results of the randomized multicenter BONAMI trial. Eur Heart J 32(14):1748–1757. doi:ehq455 [pii] 10.1093/eurheartj/ehq455

Sato T, Iso Y, Uyama T, Kawachi K, Wakabayashi K, Omori Y, Soda T, Shoji M, Koba S, Yokoyama S, Fukuda N, Saito S, Katagiri T, Kobayashi Y, Takeyama Y, Umezawa A, Suzuki H (2011) Coronary vein infusion of multipotent stromal cells from bone marrow preserves cardiac function in swine ischemic cardiomyopathy via enhanced neovascularization. Lab Invest 91(4):553–564. doi:labinvest2010202 [pii] 10.1038/labinvest.2010.202

Schachinger V, Erbs S, Elsasser A, Haberbosch W, Hambrecht R, Holschermann H, Yu J, Corti R, Mathey DG, Hamm CW, Suselbeck T, Assmus B, Tonn T, Dimmeler S, Zeiher AM (2006a) Intracoronary bone marrow-derived progenitor cells in acute myocardial infarction. N Engl J Med 355(12):1210–1221

Schachinger V, Erbs S, Elsasser A, Haberbosch W, Hambrecht R, Holschermann H, Yu J, Corti R, Mathey DG, Hamm CW, Suselbeck T, Werner N, Haase J, Neuzner J, Germing A, Mark B, Assmus B, Tonn T, Dimmeler S, Zeiher AM (2006b) Improved clinical outcome after intracoronary administration of bone-marrow-derived progenitor cells in acute myocardial infarction: final 1-year results of the REPAIR-AMI trial. Eur Heart J 27(23):2775–2783

Schaefer A, Meyer GP, Fuchs M, Klein G, Kaplan M, Wollert KC, Drexler H (2006) Impact of intracoronary bone marrow cell transfer on diastolic function in patients after acute myocardial infarction: results from the BOOST trial. Eur Heart J 27(8):929–935. doi:ehi817 [pii] 10.1093/eurheartj/ehi817

Schomig K, Busch G, Steppich B, Sepp D, Kaufmann J, Stein A, Schomig A, Ott I (2006) Interleukin-8 is associated with circulating CD133+ progenitor cells in acute myocardial infarction. Eur Heart J 27(9):1032–1037. doi:10.1093/eurheartj/ehi761

Schuleri KH, Amado LC, Boyle AJ, Centola M, Saliaris AP, Gutman MR, Hatzistergos KE, Oskouei BN, Zimmet JM, Young RG, Heldman AW, Lardo AC, Hare JM (2008) Early improvement in cardiac tissue perfusion due to mesenchymal stem cells. Am J Physiol Heart Circ Physiol 294(5):H2002–2011. doi:00762.2007 [pii] 10.1152/ajpheart.00762.2007

Schuleri KH, Feigenbaum GS, Centola M, Weiss ES, Zimmet JM, Turney J, Kellner J, Zviman MM, Hatzistergos KE, Detrick B, Conte JV, McNiece I, Steenbergen C, Lardo AC, Hare JM (2009) Autologous mesenchymal stem cells produce reverse remodelling in chronic ischaemic cardiomyopathy. Eur Heart J 30(22):2722–2732. doi:ehp265 [pii] 10.1093/eurheartj/ehp265

Seeger FH, Tonn T, Krzossok N, Zeiher AM, Dimmeler S (2007) Cell isolation procedures matter: a comparison of different isolation protocols of bone marrow mononuclear cells used for cell therapy in patients with acute myocardial infarction. Eur Heart J 28(6):766–772. doi:ehl509 [pii] 10.1093/eurheartj/ehl509

Silva GV, Fernandes MR, Cardoso CO, Sanz RR, Oliveira EM, Jimenez-Quevedo P, Lopez J, Angeli FS, Zheng Y, Willerson JT, Perin EC (2011) A dosing study of bone marrow mononuclear cells for transendocardial injection in a pig model of chronic ischemic heart disease. Tex Heart Inst J 38(3):219–224

Siminiak T, Fiszer D, Jerzykowska O, Grygielska B, Rozwadowska N, Kalmucki P, Kurpisz M (2005) Percutaneous trans-coronary-venous transplantation of autologous skeletal myoblasts in the treatment of post-infarction myocardial contractility impairment: the POZNAN trial. Eur Heart J 26(12):1188–1195. doi:ehi159 [pii]10.1093/eurheartj/ehi159

Skali H, Zornoff LA, Pfeffer MA, Arnold MO, Lamas GA, Moye LA, Plappert T, Rouleau JL, Sussex BA, St John Sutton M, Braunwald E, Solomon SD (2005) Prognostic use of echocardiography 1 year after a myocardial infarction. Am Heart J 150(4):743–749. doi:S0002-8703(04)00874-9 [pii]10.1016/j.ahj.2004.10.033

Smith RR, Barile L, Cho HC, Leppo MK, Hare JM, Messina E, Giacomello A, Abraham MR, Marban E (2007) Regenerative potential of cardiosphere-derived cells expanded from percutaneous endomyocardial biopsy specimens. Circulation 115(7):896–908. doi:CIRCULATIONAHA.106.655209 [pii]10.1161/CIRCULATIONAHA.106.655209

Solomon SD, Skali H, Bourgoun M, Fang J, Ghali JK, Martelet M, Wojciechowski D, Ansmite B, Skards J, Laks T, Henry D, Packer M, Pfeffer MA (2005) Effect of angiotensin-converting enzyme or vasopeptidase inhibition on ventricular size and function in patients with heart failure: the Omapatrilat Versus Enalapril Randomized Trial of Utility in Reducing Events (OVERTURE) echocardiographic study. Am Heart J 150(2):257–262. doi:S0002-8703(05)00035-9 [pii] 10.1016/j.ahj.2004.09.056

St John Sutton M, Pfeffer MA, Moye L, Plappert T, Rouleau JL, Lamas G, Rouleau J, Parker JO, Arnold MO, Sussex B, Braunwald E (1997) Cardiovascular death and left ventricular remodeling two years after myocardial infarction: baseline predictors and impact of long-term use of captopril: information from the Survival and Ventricular Enlargement (SAVE) trial. Circulation 96(10):3294–3299

Staat P, Rioufol G, Piot C, Cottin Y, Cung TT, L'Huillier I, Aupetit JF, Bonnefoy E, Finet G, Andre-Fouet X, Ovize M (2005) Postconditioning the human heart. Circulation 112(14):2143–2148. doi:CIRCULATIONAHA.105.558122 [pii] 10.1161/CIRCULATIONAHA.105.558122

Strauer BE, Brehm M, Zeus T, Bartsch T, Schannwell C, Antke C, Sorg RV, Kogler G, Wernet P, Muller HW, Kostering M (2005) Regeneration of human infarcted heart muscle by intracoronary autologous bone marrow cell transplantation in chronic coronary artery disease: the IACT Study. J Am Coll Cardiol 46(9):1651–1658. doi:S0735-1097(05)01842-5 [pii] 10.1016/j.jacc.2005.01.069

Strauer BE, Yousef M, Schannwell CM (2010) The acute and long-term effects of intracoronary Stem cell Transplantation in 191 patients with chronic heARt failure: the STAR-heart study. Eur J Heart Fail 12(7):721–729. doi:hfq095 [pii] 10.1093/eurjhf/hfq095

Takahashi K, Tanabe K, Ohnuki M, Narita M, Ichisaka T, Tomoda K, Yamanaka S (2007) Induction of pluripotent stem cells from adult human fibroblasts by defined factors. Cell 131(5):861–872. doi:S0092-8674(07)01471-7 [pii] 10.1016/j.cell.2007.11.019

Tang YL, Zhu W, Cheng M, Chen L, Zhang J, Sun T, Kishore R, Phillips MI, Losordo DW, Qin G (2009) Hypoxic preconditioning enhances the benefit of cardiac progenitor cell therapy for treatment of myocardial infarction by inducing CXCR4 expression. Circ Res 104(10):1209–1216. doi:CIRCRESAHA.109.197723 [pii] 10.1161/CIRCRESAHA.109.197723

Taylor DA, Atkins BZ, Hungspreugs P, Jones TR, Reedy MC, Hutcheson KA, Glower DD, Kraus WE (1998) Regenerating functional myocardium: improved performance after skeletal myoblast transplantation. Nat Med 4(8):929–933

Tendera M, Wojakowski W, Ruzyllo W, Chojnowska L, Kepka C, Tracz W, Musialek P, Piwowarska W, Nessler J, Buszman P, Grajek S, Breborowicz P, Majka M, Ratajczak MZ (2009) Intracoronary infusion of bone marrow-derived selected CD34+ CXCR4+ cells and non-selected mononuclear cells in patients with acute STEMI and reduced left ventricular ejection fraction: results of randomized, multicentre myocardial regeneration by intracoronary infusion of Selected Population of Stem Cells in Acute Myocardial Infarction (REGENT) Trial. Eur Heart J 30(11):1313–1321. doi:ehp073 [pii] 10.1093/eurheartj/ehp073

Theiss HD, Brenner C, Engelmann MG, Zaruba MM, Huber B, Henschel V, Mansmann U, Wintersperger B, Reiser M, Steinbeck G, Franz WM (2010) Safety and efficacy of SITAgliptin plus GRanulocyte-colony-stimulating factor in patients suffering from Acute Myocardial Infarction (SITAGRAMI-Trial)–rationale, design and first interim analysis. Int J Cardiol 145(2):282–284. doi:S0167-5273(09)01501-0 [pii] 10.1016/j.ijcard.2009.09.555

Thibault H, Piot C, Staat P, Bontemps L, Sportouch C, Rioufol G, Cung TT, Bonnefoy E, Angoulvant D, Aupetit JF, Finet G, Andre-Fouet X, Macia JC, Raczka F, Rossi R, Itti R, Kirkorian G, Derumeaux G, Ovize M (2008) Long-term benefit of postconditioning. Circulation 117(8):1037–1044. doi:CIRCULATIONAHA.107.729780 [pii] 10.1161/CIRCULATIONAHA.107.729780

Tjandrawidjaja MC, Fu Y, Westerhout CM, Wagner GS, Granger CB, Armstrong PW (2010) Usefulness of the QRS score as a strong prognostic marker in patients discharged after undergoing primary percutaneous coronary intervention for ST-segment elevation myocardial infarction. Am J Cardiol 106(5):630–634. doi:S0002-9149(10)00949-5 [pii]10.1016/j.amjcard.2010.04.013

Tjwa M, Janssens S, Carmeliet P (2008) Plasmin therapy enhances mobilization of HPCs after G-CSF. Blood 112(10):4048–4050. doi:10.1182/blood-2008-07-166587

Ulloa-Montoya F, Kidder BL, Pauwelyn KA, Chase LG, Luttun A, Crabbe A, Geraerts M, Sharov AA, Piao Y, Ko MS, Hu WS, Verfaillie CM (2007) Comparative transcriptome analysis of embryonic and adult stem cells with extended and limited differentiation capacity. Genome Biol 8(8):R163. doi:gb-2007-8-8-r163 [pii]10.1186/gb-2007-8-8-r163

Valgimigli M, Rigolin GM, Cittanti C, Malagutti P, Curello S, Percoco G, Bugli AM, Della Porta M, Bragotti LZ, Ansani L, Mauro E, Lanfranchi A, Giganti M, Feggi L, Castoldi G, Ferrari R (2005) Use of granulocyte-colony stimulating factor during acute myocardial infarction to enhance bone marrow stem cell mobilization in humans: clinical and angiographic safety profile. Eur Heart J 26(18):1838–1845. doi:ehi289 [pii]

Van't Hof W, Mal N, Huang Y, Zhang M, Popovic Z, Forudi F, Deans R, Penn MS (2007) Direct delivery of syngeneic and allogeneic large-scale expanded multipotent adult progenitor cells

improves cardiac function after myocardial infarct. Cytotherapy 9(5):477–487. doi:780539472 [pii] 10.1080/14653240701452065
Vasa M, Fichtlscherer S, Aicher A, Adler K, Urbich C, Martin H, Zeiher AM, Dimmeler S (2001) Number and migratory activity of circulating endothelial progenitor cells inversely correlate with risk factors for coronary artery disease. Circ Res 89(1):E1–7
Velagaleti RS, Pencina MJ, Murabito JM, Wang TJ, Parikh NI, D'Agostino RB, Levy D, Kannel WB, Vasan RS (2008) Long-term trends in the incidence of heart failure after myocardial infarction. Circulation 118(20):2057–2062. doi:CIRCULATIONAHA.108.784215 [pii] 10.1161/CIRCULATIONAHA.108.784215
Wagers AJ, Weissman IL (2004) Plasticity of adult stem cells. Cell 116:639–648
White HD, Norris RM, Brown MA, Brandt PW, Whitlock RM, Wild CJ (1987) Left ventricular end-systolic volume as the major determinant of survival after recovery from myocardial infarction. Circulation 76(1):44–51
Wills AA, Holdway JE, Major RJ, Poss KD (2008) Regulated addition of new myocardial and epicardial cells fosters homeostatic cardiac growth and maintenance in adult zebrafish. Development 135(1):183–192. doi:dev.010363 [pii] 10.1242/dev.010363
Wojakowski W, Tendera M, Michalowska A, Majka M, Kucia M, Maslankiewicz K, Wyderka R, Ochala A, Ratajczak MZ (2004) Mobilization of CD34/CXCR4+, CD34/CD117+, c-met+ stem cells, and mononuclear cells expressing early cardiac, muscle, and endothelial markers into peripheral blood in patients with acute myocardial infarction. Circulation 110(20):3213–3220. doi:01.CIR.0000147609.39780.02 [pii] 10.1161/01.CIR.0000147609.39780.02
Wojakowski W, Tendera M, Kucia M, Zuba-Surma E, Paczkowska E, Ciosek J, Halasa M, Krol M, Kazmierski M, Buszman P, Ochala A, Ratajczak J, Machalinski B, Ratajczak MZ (2009) Mobilization of bone marrow-derived Oct-4+ SSEA-4+ very small embryonic-like stem cells in patients with acute myocardial infarction. J Am Coll Cardiol 53(1):1–9. doi:S0735-1097(08)03343-3 [pii] 10.1016/j.jacc.2008.09.029
Wollert KC, Meyer GP, Lotz J, Ringes-Lichtenberg S, Lippolt P, Breidenbach C, Fichtner S, Korte T, Hornig B, Messinger D, Arseniev L, Hertenstein B, Ganser A, Drexler H (2004) Intracoronary autologous bone-marrow cell transfer after myocardial infarction: the BOOST randomised controlled clinical trial. Lancet 364(9429):141–148. doi:10.1016/S0140-6736(04)16626-9 S0140673604166269 [pii]
Wright DE, Bowman EP, Wagers AJ, Butcher EC, Weissman IL (2002) Hematopoietic stem cells are uniquely selective in their migratory response to chemokines. J Exp Med 195 (9):1145–1154
Zaruba MM, Soonpaa M, Reuter S, Field LJ (2010) Cardiomyogenic potential of C-kit(+)-expressing cells derived from neonatal and adult mouse hearts. Circulation 121(18):1992–2000. doi:CIRCULATIONAHA.109.909093 [pii] 10.1161/CIRCULATIONAHA.109.909093
Zeng L, Rahrmann E, Hu Q, Lund T, Sandquist L, Felten M, O'Brien TD, Zhang J, Verfaillie C (2006) Multipotent adult progenitor cells from swine bone marrow. Stem Cells 24(11):2355–2366. doi:2005-0551 [pii] 10.1634/stemcells.2005-0551
Zeng L, Hu Q, Wang X, Mansoor A, Lee J, Feygin J, Zhang G, Suntharalingam P, Boozer S, Mhashilkar A, Panetta CJ, Swingen C, Deans R, From AH, Bache RJ, Verfaillie CM, Zhang J (2007) Bioenergetic and functional consequences of bone marrow-derived multipotent progenitor cell transplantation in hearts with postinfarction left ventricular remodeling. Circulation 115(14):1866–1875. doi:CIRCULATIONAHA.106.659730 [pii] 10.1161/CIRCULATIONAHA.106.659730
Zohlnhofer D, Ott I, Mehilli J, Schomig K, Michalk F, Ibrahim T, Meisetschlager G, von Wedel J, Bollwein H, Seyfarth M, Dirschinger J, Schmitt C, Schwaiger M, Kastrati A, Schomig A (2006) Stem cell mobilization by granulocyte colony-stimulating factor in patients with acute myocardial infarction: a randomized controlled trial. JAMA 295(9):1003–1010. doi:295/9/1003 [pii] 10.1001/jama.295.9.1003
Zohlnhofer D, Dibra A, Koppara T, de Waha A, Ripa RS, Kastrup J, Valgimigli M, Schomig A, Kastrati A (2008) Stem cell mobilization by granulocyte colony-stimulating factor for myocardial recovery after acute myocardial infarction: a meta-analysis. J Am Coll Cardiol 51(15):1429–1437. doi:S0735-1097(08)00385-9 [pii] 10.1016/j.jacc.2007.11.073

Chapter 8
Clinical Gene and Stem Cell Therapy in Patients with Acute and Chronic Myocardial Ischemia

Jens Kastrup, Annette Ekblond, Mandana Haack-Sørensen, Anders B. Mathiasen and Abbas A. Qayyum

Abstract The clinical consequence of the increasing incidence of coronary artery disease is a growing problem worldwide and ischemic heart disease remains the most common cause of death and a major cause of hospital admissions in industrialised countries. Early treatment with stabilizing drugs and mechanical revascularization by percutaneous coronary intervention or coronary by-pass surgery has reduced the mortality significantly. However, there is still a group of patients, which cannot be treated satisfactorily with these interventions. Treatment with genes encoding vascular growth factors or stem cells with the potential to regenerate the damaged myocardium is a relatively new approach. The results from early clinical studies on gene and stem cell therapy for cardiac regeneration in patients with acute or chronic ischemic heart disease have been inconsistent. Some of the discrepancy could be due to differences in study designs or patient selection. This review will present the conducted clinical trials and try to clarify the influence of patient selection, chosen cell type, cell source, delivery methods and mechanisms of action on the differences in results.

Introduction

Ischemic heart disease (IHD) caused due to coronary artery disease (CAD) remains a major cause of mortality and morbidity in industrialized countries. Early treatment with stabilizing drugs and percutaneous coronary intervention (PCI) or coronary bypass surgery (CABG) has reduced morbidity and mortality significantly. Morbidity in survivors of the acute coronary syndrome is related to myocardial remodelling and progressive heart failure. It has, therefore, been a challenge in clinical research to find new treatment modalities that aim to reduce and repair the myocardial damage and improve blood supply to the myocardium in the ischemic heart.

J. Kastrup (✉) · A. Ekblond · M. Haack-Sørensen · A. B. Mathiasen · A. A. Qayyum
Cardiology Stem Cell Centre and Cardiac Catheterization Laboratory 2014,
The Heart Centre, Rigshospitalet Copenhagen University Hospital and Faculty of Health Sciences, Blegdamsvej 9, 2100 Copenhagen, Denmark
e-mail: jens.kastrup@regionh.dk

J. Hescheler, E. Hofer (eds.), *Adult and Pluripotent Stem Cells*,
DOI 10.1007/978-94-017-8657-7_8,

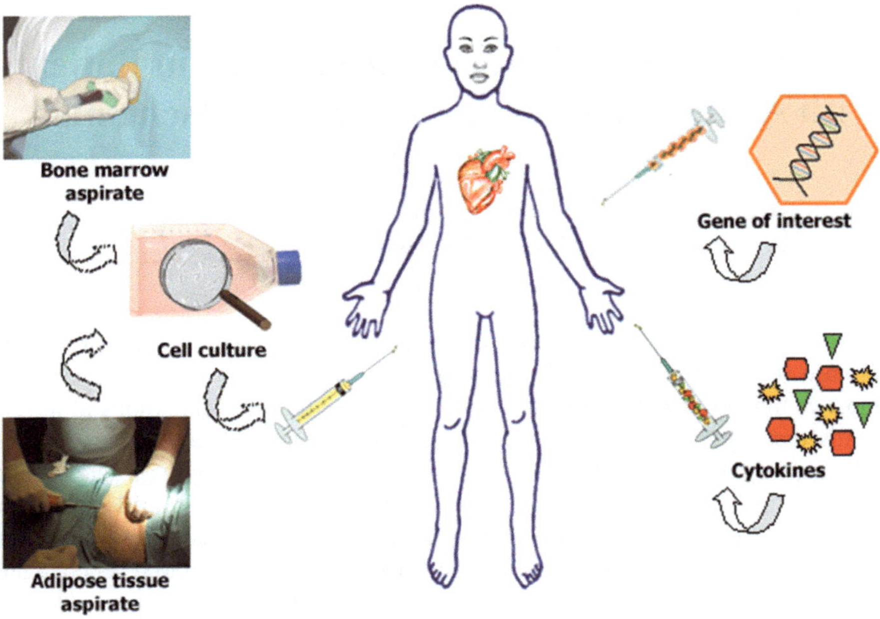

Fig. 8.1 Patients with ischemic heart disease can be treated with either cytokines, genes encoding cytokines or stem cells obtained from different tissues

The biological consequences of CAD are reduced blood flow to the myocardium with hypoxia during stress and rest conditions. The clinical presentation of CAD is acute myocardial infarction (MI) and chronic ischemic heart disease (CIHD) caused by occlusion or narrowing of a coronary artery, respectively. The clinical consequences differ between the acute and chronic manifestations. In MI there is an acute loss of cardiomyocytes and scar formation, contra-lateral hypertrophy, resulting in end stage dilated cardiomyopathy and heart failure (IHF). The CIHD is characterized by an adverse remodeling of the heart with cardiomyocyte hypertrophy, increased fibrosis and capillary rarification (Hinkel et al. 2011).

The regenerative capacity within the heart to repair the consequences of IHD is very poor. There is some regenerative capacity, albeit insufficient to offset the cell death in heart disease (Willems et al. 2011). The angiogenesis and cardiomyogenesis are not able to repair and restore cardiac function. Therefore, regenerative treatment with genes encoding vascular growth factors and stem cells have had increasing interest in recent years and have been evaluated pre-clinically and in clinical trials in patients with acute and chronic IHD (Ahn et al. 2008a, b; Kastrup 2011; Mathiasen et al. 2009; Fig. 8.1). The treatment intensions were to induce growth of new blood vessels and/or replacement of damaged myocardial cells either directly by trans-differentiation of stem cells or by a paracrine effect of cytokines secreted from the stem cells or transfected cells.

The aim of this review is to discuss current issues on gene and stem cell therapy in patients with ischemic heart disease with focus on conducted clinical trials and the perspective for these treatments in the future.

Table 8.1 Growth factors—adhesion molecules—enzymes etc. with the potential to induce angiogenesis in ischemic tissue

VEGF	Vascular endothelial growth factor- A, B, C, D, E
FGF	Fibroblast growth factor-1, 2, 4, 5
Ang1, 2	Angiopoietin-1, 2
HIF-1-α	Hypoxia inducible factor-1-α
TGF-α, β	Transforming growth factor-α, β
G-CSF	Granulocyte colony-stimulating factor
GM-CSF	Granulocyte-macrophage colony-stimulating factor
IGF1	Insuline growth factor 1
HGF	Hepatocyte growth factor
MCP1	Monocyte chemotactic protein-1
PDGF	Platelet-derived growth factor-BB
eNOS	endothelial-nitric-oxide synthase
iNOS	inducible-nitric-oxide synthase
VE-Cadherin	Vascular Endothelial-Cadherin

Biological Regeneration

Angiogenesis/Vasculogenesis

The formation of new blood vessels is inherent in vascular tissue, and it can be induced by trauma, ischemia, and inflammation or tumor growth (Ahn et al. 2008a, b). To create new blood vessels it is necessary to establish an interaction between locally produced cytokines and cells derived from the tissue area and the blood circulation (Ahn et al. 2008a, b). Local tissue ischemia due to a decrease in blood flow and reduction in necessary substrates such as oxygen has been shown to stimulate angiogenesis and vasculogenesis in many experimental models. This angiogenesis can be initiated by local production and secretion of vascular growth factors and by stem cells. In animal studies it has been demonstrated that several families of proteins have an angiogenic potential in ischemic tissues (Ahn et al. 2008a, b; Table 8.1). However, only a few of these vascular growth factors such as vascular endothelial growth factor (VEGF) and fibroblast growth factor (FGF) have moved from pre-clinical animal studies into clinical trials.

Cardiomyogenesis

The heart has a reduced capacity to regenerate cardiomyocytes after an acute myocardial infarction. It is now evident that new cardiomyocytes can arise in the heart. Several different potential cell sources have been discovered; a subpopulation of mononuclear cardiomyocytes, circulating progenitor cells derived from the bone marrow, epicardial cells and progenitor cells in the heart have been identified with the potential to differentiate into cardiomyocytes with the proper stimulation (Hinkel et al. 2011). However, the regenerative capacity of these cell populations does not have the capacity to restore destroyed cardiac tissue and function.

Gene Therapy for Clinical Studies

The purpose of gene therapy is to induce a local production of growth factors for a prolonged period of time and hereby produce a steady biological effect (Ahn et al. 2008a, b). The gene encoding the growth factor is delivered to the cells as a cDNA (complementary DNA) formulation, which is then transcribed into the nucleus of the cell (transfection). Then the production of the growth factor protein is initiated. The genes can be delivered to the myocardium either by intracoronary (IC) infusions or intramyocardial (IM) injections during by-pass surgery or using a less traumatic percutaneous method (Baldazzi et al. 2008; Kastrup et al. 2005). In theory the IM method will cause more growth factor to remain in the heart and thus reducing the potential side effects caused by increased systemic levels of growth factors in non-ischemic tissue outside the treatment area.

Two different methods have been used for gene transfection: the naked plasmid-DNA and viral vectors (adeno-associated virus or adenovirus). Plasmid-DNA transfection is rather simple, but the efficacy is low. It is estimated that only about 1 % of the plasmid DNA is entering the cells (Ahn et al. 2008a, b). Adenovirus enters the cells using specific cell surface receptors. However, it is a problem that the adenovirus induces immunological and inflammatory reactions, which can reduce the period of gene activity and inhibit a later re-administration of the adenovirus. With plasmid/adenovirus-DNA it is estimated that there will only be approximately four weeks of growth factor production, and then the genes are metabolised and removed from the cells.

In patients with IHD clinical gene therapy trials have been conducted with plasmid and adenovirus formulations encoding the growth factors VEGF or FGF (Tables 8.2 and 8.3). The study population have in almost all studies been patients with severe coronary artery disease, which could not be treated optimally with conventional revascularization therapies. The genes were initially delivered to the myocardium either by intracoronary (IC) or intramyocardial (IM) injections during CABG or with a thoracotomy alone. In recent years most studies have been conducted with IM injections using a less traumatic percutaneous delivery systems such as the NOGA system (Biological Delivery System®, Cordis, US).

Clinical Intracoronary Gene Trials

Intracoronary injections of plasmid/liposome-VEGF-A_{165} and Ad-VEGF-A_{165} have been compared in patients undergoing elective PCI for coronary artery stenosis (Hedman et al. 2003; Laitinen et al. 2000; Table 8.2). The myocardial perfusion improved significantly in the Ad-VEGF-A_{165} treated patients but not in plasmid/liposome-VEGF-A_{165} treated patients or control patients. However, there was no difference in clinical improvements between the three groups. The results indicated that an increased transfection efficacy of adenovirus compared to plasmid might be of importance for the clinical outcome, although the trial was underpowered to test this point.

Table 8.2 Intracoronary treatment with vascular growth factors genes in chronic ischemic heart disease

	n	Growth factor	Study design	Treatment effect
VEGF-A$_{165}$				
Laitinen et al. 2000	15	P1.-VEGF-A$_{165}$1,000 μg	10:5(placebo)	Safety study. Treatment effect not measured
Hedmann et al. 2003 KAT study	103	P1.-VEGF-A$_{165}$ and Ad.-VEGF-A$_{165}$	28:37:38 (placebo) Double-blind	Improved perfusion with Ad.-VEGF-A$_{165}$
FGF4				
Grines et al. 2002 Agent-1	69	Ad-FGF43.3 × 10^8–10^{11} particle unit (pu)	9:9:9:22:11:19 (placebo) Double-blind	(Tendency to improved exercise time in subgroup)
Grines et al. 2003 Agent-2	52	Ad-FGF4 10^{10} ad5-FGF4 virus particle unit (pu)	35:17 (placebo) Double-blind	Improved perfusion
Henry et al. 2007 Agent-3 and -4	456 250	Ad-FGF4	180:1175:177 (placebo) Double-blind	No improvement exercise. Improvement in exercise in women.

Several clinical trials have been conducted with the adenovirus FGF4 gene (ad5-FGF4) in patients with IHD with less severe CAD than in most other clinical gene therapy trials (Grines et al. 2002; Grines et al. 2003). The patients were all evaluated by a core angiography laboratory to be suitable candidates for PCI or CABG. The AGENT-1 trial randomised 69 patients in a double-blind placebo-controlled design to 6 ascending doses of ad5-FGF4 virus into all major coronary arteries (Grines et al. 2002). The infusion was generally well tolerated, but a majority of the patients had a rise in neutralizing antibodies to the adenovirus. There were no significant improvements in the primary endpoint; exercise capacity, neither for the individual doses nor for the pooled data of all the doses used. In a subgroup analysis, an improvement was discovered in patients with baseline exercise time less than 10 min.

The follow-up AGENT 2 study was also a double-blind placebo-controlled design (Grines et al. 2003). This trial including 37 patients demonstrated improved myocardial perfusion compared to the placebo group. Then the two larger multicentre studies AGENT-3 and 4 were initiated, aiming to include a total of 456 and 250 patients, respectively. However, these two trials were both terminated prematurely January 2004 after an interim analysis in AGENT-3 demonstrated lack of efficacy. The post-hoc combined analysis of the AGENT-3 and 4 could not detect any effect of the treatment on exercise or clinical parameters. It is however interesting that when dividing the patients into genders, a clear difference in response between men and women was demonstrated. There was a clear treatment effect of Ad5FGF-4 in women, but not in men. Women had increased total exercise time, time to 1 mm ST-depression, time to angina and a reduction in CCS class (Henry et al. 2007).

Clinical Intra-Myocardial Gene Trials

The first study was an un-blinded evaluation of two different doses of plasmid VEGF-A_{165} injected IM via a mini-thoracotomy in 20 patients with CAD in general anaesthesia without any problems (Symes et al. 1999; Table 8.3). The study found an improvement in myocardial perfusion at follow-up and the treatment was safe. Then VEGF-A_{165} gene therapy IM was compared with direct myocardial revascularization with laser therapy (DMR) (Tio et al. 2004). The myocardial perfusion was improved in the VEGF-A_{165} group in the regional ischemic area, but not compared to DMR treated and controls.

The results from the early small studies were followed by the randomized double-blind placebo-controlled EUROINJECT-1 trial, which was the first larger gene therapy study using the percutaneous delivery technique (Baldazzi et al. 2008; Kastrup et al. 2005). A total of 40 patients with chronic CIHD received plasmid VEGF-A_{165}injections and 40 patients received placebo plasmid injections into an ischemic region of the left ventricle. Myocardial perfusion improved in both the VEGF-A_{165} group and the placebo group, but there were no statistically significant differences between the groups. The local linear myocardial shortening of the treated region measured with the NOGA method was significantly higher in VEGF-A_{165} treated patients compared to placebo (Gyongyosi et al. 2005). The nitroglycerine consumption was also reduced significantly in the gene transfer group compared to the placebo group. The lack of significant effect on myocardial perfusion of the treatment could potentially be explained by a too low VEGF-A_{165}dose or a too low transfection rate. However, in a recent published study using a four times higher plasmid VEGF-A_{165} dose (2,000 μg), Stewart et al. (2009) could not detect any effect of the treatment on any of the endpoints in the study (Stewart et al. 2009). There was no difference between the VEGF-treated and the placebo groups in the primary end point of change in myocardial perfusion from baseline to 3 or 6 months, assessed by single photon emission computed tomography (SPECT), although a significant reduction in the ischemic area was seen in both groups. Also, similar improvements in exercise capacity and angina symptoms were seen in the VEGF and the placebo groups at 3 and 6 months, although again there were no differences between these groups. This indicates that a more effective transfection rate using i. e. adenovirus, transfected cells or combined gene and stem cell therapy could be needed to induce neovascularisation in the ischemic myocardium.

Using the percutaneous NOGA® method adenovirus-5-VEGF-121 (AdV5-121) was investigated in a small safety and feasibility study in 10 patients with stable refractory angina (Fuchs et al. 2006). The treatment was safe, but no clear improvement in clinical data was seen. In the REVASC study 67 patients were randomized to either maximum medical therapy or AdVEGF-121 IM injections via a minithoracotomy (Stewart et al. 2006). AdVEGF-121 treated had significantly improved exercise capacity and angina symptoms at weeks 12 and 26. However, there was no effect on myocardial perfusion measured by SPECT. There were some procedure related adverse events in the thoracotomy group, but there was no overall significant difference in adverse events between the groups.

Table 8.3 Intra-myocardial delivery of vascular growth factors genes in chronic ischemic heart disease

	n	Growth factor dose	Study design	Treatment effect
Pl.-VEGF-A$_{165}$/FGF				
Symes et al. 1999	20	125 μg (10):250 μg (10)	10:10:0 (controls)	Improved perfusion on SPECT and angiography
Tio et al. 2004	10	500 μg	10:12 DMR[a]: 13 (controls)	Improved perfusion on PET with VEGF
Kastrup et al. 2005 EUROINJECT trial	80	500 μg	40:40 (placebo) Double-blind	No improvement in perfusion. Improved wall motion
Ripa et al. 2006	48	500 μg and ug/kg body weight G-CSF for 6 days	16 VEGF + G-CSF: 10 16 VEGF: 16 (placebo)	No improvement in perfusion on SPECT or in clinical data
Stewart et al. 2009 Northern	93	2,000 μg	48:45 (placebo) Double-blind	No improvement in perfusion or clinical data
Pl.-VEGF-A$_{165}$/FGF				
Kukula et al. 2011 VIF-CAD	52	500 μg	33:19 (placebo) Double-blind	No improvement in perfusion on SPECT. Improved exercise tolerance and clinical symptoms
Ad-VEGF-A$_{121}$				
Fuchs et al. 2006	10	Ad.-VEGF-A$_{121}$	6:4 (placebo) Double-blind	Treatment safe. No improvement in clinical data
Stewart et al. 2006 REVASC	67	Ad.-VEGF-A$_{121}$ 4×10^{10} particle unit	32:35 (controls)[b]	Increased exercise time
Kastrup 2010 NOVA	17	Ad.-VEGF-A$_{121}$ 4×10^{10} particle unit	12:5 (placebo) Double-blind	Treatment safe. No improvement in clinical data
Pl.-VEGF2				
Losordo et al. 2002	19	200 μg (6):800 μg (6):2000 μg (1)	12:7 (placebo)	Increased exercise time and improved symptoms
Genasis 2006 Unpublished	295 (404 lanne)	–	–	No improvement in exercise at interim analysis. Study stopped

Pl.-Plasmid, *Ad.*-adenovirus. + Treatment effect, - No treatment effect. *I.m.* Intra-myocardial, *I.c.* Intracoronary, *I.v.* Intravenous,
[a] Direct myocardial revascularization (*DMR*) with laser therapy. *SPECT*–Single Photon Emission Computerized Tomography
[b] The controls crossed over to active treatment 90 days after inclusion in trial

The NOVA trial was then designed as a double-blind placebo-controlled study of the safety and efficacy of BIOBYPASS (Ad_{GV}VEGF121.10NH) gene therapy in 129 patients with refractory advanced CAD. The study was initiated April 2005 and terminated prematurely February 2007 due to slow inclusion rate, which led to a company product portfolio decision by the company. A total of 17 patients were included in the study. It was demonstrated, that IM injection of BIOBYPASS was safe, but it did not improve exercise capacity, time to ischemic threshold or myocardial perfusion compared to sham injection in patients with refractory myocardial ischemia (Kastrup et al. 2011).

Based on these results of a phase I/II placebo-controlled double-blind dose escalating plasmid VEGF-2 trial was designed (Losordo et al. 2002; Table 8.3). The study was scheduled to include nine patients in each dose treatment group. However, after inclusion of 19 patients, the trial was interrupted by the FDA in the wake of the death of an 18-years-old subject enrolled in an unrelated study of ornithine transcarbamylase deficiency involving adenoviral GTx. Accordingly; data for the used doses of VEGF-2 were pooled to improve statistical power. At 12 weeks follow-up, angina pectoris was significantly reduced in all active treated patients compared to placebo treated. There was a non-significant trend towards improvement in exercise capacity and the Seattle Angina Questionnaire. Moreover, the larger GENASIS trial planned to include 404 patients with CIHD for VEGF2 plasmid or placebo treatment was stopped after an interim analysis of 295 patients [Unpublished]) (Table 8.3).

Combined Clinical Genes or Gene/Stem Cell Trials

A recent published study, the VIF-CAD trial, investigated the combination of VEGF-A_{165}/bFGF IM injections in patients with CIHD in a double-blind placebo-controlled trial (Kukula et al. 2011; Table 8.3). There was no change in myocardial perfusion, but exercise capacity and symptoms improved.

The combination of treatment with VEGF-A_{165} gene and G-CSF mobilization of stem cells was evaluated in 16 patients with stable CIHD. Patients were treated with IM injections of VEGF-A_{165} plasmid followed one week later by daily subcutaneous injection of 10 μg/kg body weight G-CSF (Neupogen®) for six days (Ripa et al. 2006). The treatment was compared with two control groups; I) 16 patients treated with VEGF gene transfer alone and II) 16 patients treated with placebo gene injections. The combined VEGF-$A_{165}$and G-CSF treated group could not demonstrate any changes in myocardial perfusion at rest or stress between baseline and follow-up. Moreover, there were no improvements in ventricular function, exercise capacity or symptoms in any of the groups (Ripa et al. 2006; Wang et al. 2005).

Stem Cells for Clinical Treatment

Within cardiology most focus has been on bone marrow derived stem cells. Several clinical safety and efficacy studies with mononuclear cell solutions (MNC) and selected endothelial progenitor cells (EPC) from the bone marrow have been conducted in patients with acute myocardial infarction (MI), CIHD or ischemic heart failure (IHF). The results have been conflicting. Some but not all studies suggested a beneficial effect on myocardial perfusion, function and symptoms in both acute and chronic myocardial ischemia (Janssens et al. 2006; Lunde et al. 2006; Perin et al. 2003; Schachinger et al. 2006a). The variation in clinical responses of MNC treatment could be due to the fact that the MNC solution from the bone marrow is a mixture of different cell types and the percentage of therapeutically useful stem cells is very small (2 %). However, it is still debatable whether the potential clinical effect is caused directly by the stem cells or indirectly by cytokines secreted from the MNCs (Kastrup 2011).

Based on these conflicting results, there has been a request for larger randomized placebo-controlled studies. Moreover, the research has changed towards the use of more specific defined stem cell lines. Focus has moved towards more specific stem cell subpopulations such as the multi-potent mesenchymal stromal cell (MSC) from the bone marrow or the adipose derived stem cell (ADSC). MSC and ADSC can easily be isolated from the bone marrow and abdominal adipose tissue respectively, and *in vitro* culture expanded and stimulated to differentiate into different cell lines as endothelial, cardiac or other cell types (Kastrup 2010; Mathiasen et al. 2009; Pittenger and Martin 2004).

Clinical Trials with Mononuclear Cells

More and more data are accumulating with clinical MNC therapy in patients with IHD. However, the majority of experience is still from small non-randomized and a few randomized, controlled trials conducted in patients with acute MI, CIHD and IHF with either intravenous (IV) infusion, IC infusion or IM injection of stem cells either catheter guided or directly during coronary artery by-pass grafting.

Acute Myocardial Infarction

The initial open-label trials of autologous MNCs and the later larger independent randomized trials have all documented safety and feasibility of the intracoronary delivery of the cells in patients with MI (Table 8.4).

The open randomized BOOST trial in 30 MI patients (the BOOST trial) demonstrated increased left ventricle ejection fraction (LVEF) and enhanced systolic function after 6 months (Wollert et al. 2004), but after 18 and 61 months the difference in LVEF was no longer significant (Meyer et al. 2006, 2009), suggesting

Table 8.4 Intracoronary controlled trials with bone marrow mononuclear cells in patients with acute myocardial infarction

	N treated/controls	Cell type	Cell number	Treatment after MI	Treatment effect LVEF	Infarct size
BOOST 2004 [34]	30/30	MNC	2500×10^6	Day 6±1	↑(6 months)/ ↔(18+61 months)	↔
TOPCARE-AMI 2004 [42]	30/29	MNC/CPC	$213 \times 10^6/11 \times 10^6$	Day 5	↑(4 months)	↑Viability
REPAIR-AMI 2006 [43]	101/103	MNC	240×10^6	Day 3–6	↑(4 months)	NA
ASTAMI 2006 [28]	50/50	MNC	87×10^6	Day 6	↔(6 months)	↓
FINCELL 2008 [37]	40/40	MNC	360×10^6	Day 3	↑(6 months)	NA
Leuven-AMI 2009 [27]	33/34	MNC	170×10^6	Day 1	↔(4 months)	↓
REGENT 2009 [39]	80/80/40	MNC (unselected vs CD34+/ CXCR4+)	180×10^6 vs 1.9×10^6	Day 3–12	↔ between groups (6 months)	NA
BALANCE 2009 [40]	62/62	MNC	61×10^6	Day 7	↑ (60 months)	↓
MYSTAR 2009 [38]	60/0	MNC (3–6 weeks) vs MNC (3–6 months)	i.m. 200×10^6 vs 200×10^6 i.c. 1300×10^6 vs 1300×10^6	Day 21–42 vs 90–120	↑ (9–12 months)	↓
Traverse et al. 2010 [46]	30/10	MNC	100×10^6	Day 4–5	↔(6 months)	NA
Wohrle et al. 2010 [48]	29/13	MNC	381×10^6	Day 5–7	↔(6 months)	↔
HEBE 2011[36]	69/66/65	MNC/CPC/control	$296 \times 10^6/287 \times 10^6$	Day 4–7	↔(4 months)	↔
BONAMI 2011 [49]	52/49	MNC	98×10^6	Day 4–7	↔(3 months)	↑Viability

MNC Bone marrow mononuclear cells, *CPC* circulating progenitor cells, *i.c.* intracoronary; *NA* Not available that stem cell treatment merely accelerated LVEF recovery. However, the long-term data found that LVEF improved significantly in patients with transmural infarction compared to patients with non-transmural infarction. The TOPCARE-AMI trial, open randomized study, compared MNCs and circulating progenitor cells (CPC) in MI patients, and found improved LVEF and wall motion after 4 months (20 patients) (Assmus et al. 2002) and also after 1 and 5 years in a larger study (59 patients) (Leistner et al. 2011; Schachinger et al. 2004). There was no difference in effect between cell types. These results were the background for the initiation of the larger multi-center double-blinded placebo-controlled REPAIR-AMI trial with MNCs in 204 STEMI patients The study showed improved LVEF and reduced mortality and re-hospitalization after 4 months (Schachinger et al. 2006a) and the results were also stable after 1 year and 2 years although it was not designed to test this modality (Schachinger et al. 2006b). In opposition to these results, several other published trials have not been able to detect any effect of MNCs treatment on LV function after MI. The ASTAMI trial could not detect any improvement in LV function measured by magnetic resonance imaging (Erbs et al. 2005) and SPECT in a randomized trial with IC injection of MNCs in 100 patients with MI for up to three years (Beitnes et al. 2011; Lunde et al. 2006). Moreover, the Leuven MI trial with treated 67 patients with MI with either IC MNCs or placebo without any detectable effect on LV recovery measured by magnetic resonance imaging (Erbs et al. 2005; Janssens et al. 2006). However, the reduction in infarct size was greater in MNCs treated compared to placebo. The BALANCE trial included 124 patients in a MNC vs control treatment arm (Yousef et al. 2009). The study demonstrated improvement in LVEF and infarct size, which persisted for 60 months. In addition, there was also persistent improvements in exercise capacity and mortality in the long-term follow-up. In the HEBE trial, 200 patients with MI were treated with either MNC from bone marrow; CPC isolated form peripheral blood or treated with standard therapy (Hirsch et al. 2011). The study could not demonstrate any improvements in global or regional LV systolic function assessed with MRI. The same lack of effects on LV function was seen in three recently published randomized trials with MNC treatment in patients with MI (Leistner et al. 2011; Roncalli et al. 2011; Wohrle et al. 2010). The effect of repeated injections of MNCs in MI patients has also been investigated. Patients with left ventricular ejection fraction less than 45 % after MI were randomly assigned stem cell delivery via IM and IC infusion 3–6 weeks or 3–4 months after MI (Gyongyosi et al. 2009). This combined MNC delivery treatment induced a moderate but significant improvement in myocardial infarct size and left ventricular function.

In parallel with the MNC studies other research groups have focused on the use of a more specific stem cell line or progenitor cell from the bone marrow to improve the outcome of the treatment. In the multicenter study randomizing REGENT trial, Tendera et al. compared MNCs with selected cells positive for surface markers CD34 and CXCR4 in 200 patients with MI (Tendera et al. 2009). The study demonstrated no significant differences between the groups in absolute change in LVEF measured with MRI. However, this result should be interpreted with caution. The authors did observe an absolute increase of 3 % in the patients treated with either of the two cell solutions compared to no change in the control group. This difference did not reach statistical significance but is comparable to the effect of cell therapy found in the larger randomized and blinded REPAIR-AMI trial (2.5 %, $p=0.01$) (Schachinger et al. 2006b).

It is unclear why there is such an inconsistence in the results from the conducted clinical trials. The variation could be due to variations in the methods to isolate cells, in the number of cells delivered, chosen delivery methods and in timing of delivery of cells after MI as well as patient phenotype affecting MNC heterogeneity and engraftment.

A meta-analysis from 2010 including 10 randomized controlled trials of bone marrow stem cell transplantation following MI in 814 patients reported a 3.79 % improvement in LVEF compared to controls ($p<0.001$). It is interesting, that the improvement was highest in trials with most women, diabetes and age above 57 years. Whether these results still stands after inclusion of recent negative trials has to be clarified in new updated meta-analyses. Moreover, two studies have demonstrated improvements in long-term survival in MNC treated compared to controls (Schachinger et al. 2006a, b; Yousef et al. 2009).

Ischemic Heart Failure

Both skeletal myoblasts and MNCs have been evaluated in patients with IHF. IM injection of skeletal myoblasts during CABG was evaluated in 94 patients in the randomized placebo-controlled MAGIC trial (Menasche et al. 2008; Table 8.5).

At the 6 months follow-up, there was no difference in regional or global LV function between the two groups. There was a trend towards more arrhythmias in the myoblast treated patients, confirming the safety concern that had been raised in earlier, non-randomized trials. MNCs were injected IM using the percutaneous NOGA® method in 14 patients with IHF (Perin et al. 2003; Fig. 8.2). The symptoms and myocardial perfusion improved for up to 12 months compared to a control group ($n=7$). The research group has recently confirmed the results in 20 patients with IHF compared to 10 controls (Perin et al. 2011). Improved LVEF and NYHA class was also demonstrated in another study in 15 patients with IHF after IM injection of MNCs (Beeres et al. 2007). The TOPCARE-CHD trial randomized 75 patients to either IC delivered MNCs or CPC with an improvement in LVEF with MNCs but not with CPCs. Two IC infusions of MNCs within a 4 months period in 32 patients with IHF were without any effect on LVEF, but with an improvement in NYHA class (Diederichsen et al. 2008). In a large trial 191 patients with IHF were treated with either MNCs or followed in a non-randomized control group for up to five years (Strauer et al. 2010). The study demonstrated improvement in LVEF, exercise capacity, oxygen uptake and long-term mortality in the entire follow-up period.

The more specific $CD133^+$ MNC subpopulation was injected IM in18 patients undergoing CABG and compared with patients with CABG alone (Ahmadi et al. 2007). A significant improvement of regional wall motion of akinetic/dyskinetic segments and perfusion of infracted areas was demonstrated in patients receiving local

Table 8.5 Clinical trials with stem cells in patients with chronic ischemic heart failure

	N treated/ controls	Cell type	Cell number	Administration	Treatment effect LVEF	Symptoms
Perin et al. 2003	11/9	MNC	NA	i.c.	↔(12 months)	NYHA↓, exercise↑
Assmus et al. 2006	28/24/23	MNC vs CPC	$205 \times 10^6/22 \times 10^6$	i.c.	MNC↑/CPC↔ (3 months)	NYHA MNC↓/CPC↔
Beeres et al. 2007	15/0	MNC	94×10^6	i.c.	↑(3 months)	Exercise↑
Ahmadi et al. 2007	18/9	MNC		i.m. + CABG	NA	Perfusion↑, wall motion↑
Diederichsen et al. 2008	32/0	MNC (baseline) vs MNC (4 months)	$650 \times 10^6/890 \times 10^6$	i.c.	↔(12 months)	NYHA↓
Menasche et al. 2008	33/34/30	SKM	$400 \times 10^6/800 \times 10^6$	i.m.	↔(6 months)	NA
Strauer et al. 2010	191/200	MNC	66×10^6	i.c.	↑(60 months)	Exercise↑, mortality↓
Perin et al. 2011	20/10	MNC	70×10^6	i.m.	↔(6 months)	NYHA↓

MNC Bone marrow mononuclear cells, *SKM* skeletal myoblasts, *i.c.* intracoronary, *i.m.* intramyocardial, *LVEF* left ventricular ejection fraction, *NYHA* New York Heart Association class, *NA* not available

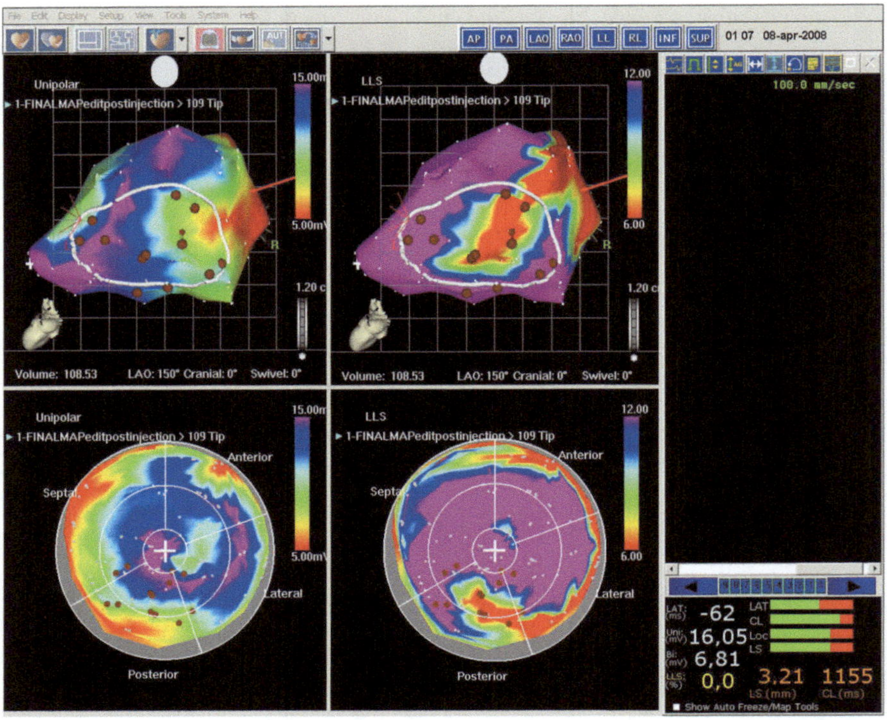

Fig. 8.2 NOGA map of left ventricle with injection of stem cells in brown spots

transplantation of CD133$^+$ cells in conjunction with CABG surgery, compared with the control group.

In conclusion, MNC treatment seems to be effective in patients with IHF, but larger clinical placebo-controlled studies are needed.

Chronic Ischemic Heart Disease

There are a small number of small open non-randomized trials using treatment with MNCs, CD34$^+$ and CPCs in patients with CIHD, (Table 8.6) LVEF, exercise capacity and symptoms improved in 20 patients with CIHD after IM injection of MNCs (Beeres et al. 2006). A few years later the same group demonstrated in 24 patients with CIHD that MNCs improved diastolic function at 3 months follow-up (Beeres et al. 2008). Kang et al. demonstrated improved LVEF after IC infusion of MNCs in an opened total coronary occlusion (Kang et al. 2006). In 27 patients with advanced CIHD IM injection of MNCs was performed and the study showed improvement of symptoms (Canadian Cardiovascular Society angina class (CCS)), exercise capacity and stress induced ischemic score after one year. These data was supported by in the PROTECT-CAD trial, where IC infusion of MNCs in 28 patients with CIHD

Table 8.6 Clinical trials with bone marrow mononuclear cells in patients with chronic ischemic heart disease

	N treated/controls	Cell type	Cell number	Administration	Treatment effect LVEF	Symptoms
Erbs et al. 2005	13/13	CPC	69×10^6	i.c.	↑ (3 months)	NA
Fuchs et al. 2006	27/0	MNC	28×10^6	i.c.	↔(3 and 12 months)	CCS↓, exercise↑
Beeres et al. 2006	20/0	MNC	$30–100 \times 10^6$	i.c.	↑(6 months)	CCS↓, exercise↑
Hendrikx et al. 2006	10/10	CABG+/-MNC	60×10^6	i.m.	↔(4 months)	NA
Kang et al. 2006	25/25 vs 16/16	MNC	1400×10^6	i.c.	MI↑/CIHD↔ (6 months)	NA
Tse et al. 2007	9/10/9	MNC	17×10^6, 42×10^6	i.m.	↑(6 months)	NYHA↓, exercise↑
Losordo et al. 2007	6/6/6/6	$CD34^+$	0.05×10^6, 0.1×10^6 or 0.5×10^6	i.m.	NA(6 months)	CCS↓, exercise↑, NTG↓
Beeres et al. 2008	24/0	MNC	85×10^6	i.c.	↔(3 months)	Diastolic function↑
Van Ramshorst et al. 2009	25/25	MNC	98×10^6	i.m.	↑(3 months)	CCS↓, perfusion↑

MNC Bone marrow mononuclear cells, *CPC* circulating progenitor cells, *i.c.* intracoronary, *i.m.* intramyocardial, *LVEF* left ventricular ejection fraction, *CCS* Canadian Cardiovascular Society angina class, *NYHA* New York Heart Association class, *NTG* Nitroglycerin consumption. *NA* = not available

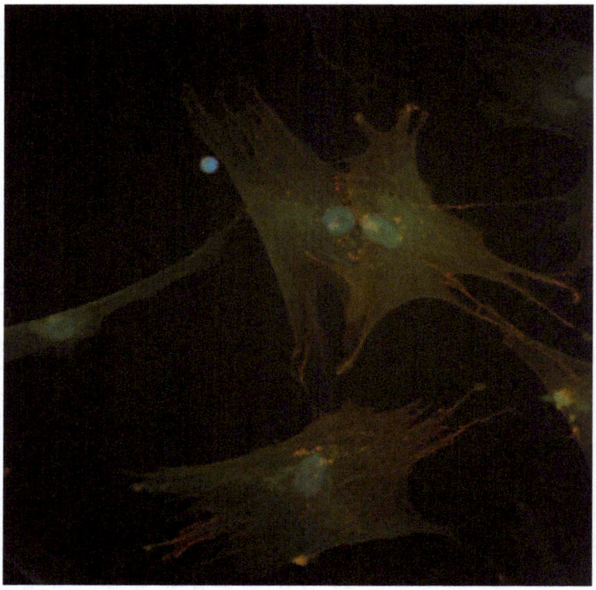

Fig. 8.3 Bone marrow derived mesenchymal stromal cell during culture expansion

improved NYHA class, exercise time and LVEF. In a randomized, double-blind, placebo-controlled trial van Ramshorst et al. treated 50 patients with IM injections of MNCs or saline (van Ramshorst et al. 2009). The study demonstrated improved myocardial perfusion, LVEF, exercise capacity and CCS class. Isolated CD34+ cells were injected IM in patients with CIHD (Losordo et al. 2007). The patients improved in angina, exercise capacity and anti-angina medication.

In conclusion, regardless of aspects of cell heterogeneity there are consistent and promising results using MNCs for the treatment of CIHD.

Clinical Trials with Mesenchymal Stromal Cells in Ischemic Heart Disease

More and more studies are coming up using the therapeutic potency of human in vitro culture-expanded autologous and allogeneic MSCs from the bone marrow in patients with IHD (Katritsis et al. 2005; Chen et al. 2004, 2006; Katritsis et al. 2007; Fig. 8.3; Table 8.7).

MSCs were in an open study administered IC in 11 MI patients without any safety concerns and the study showed improved wall motion index and non-viable segments (Katritsis et al. 2005). The arrhythmogenic side-effects of MSCs treatment were evaluated of the same group in five patients with an ICD unit (Katritsis et al. 2007). In the 16–36 months follow up period there was no episodes of either sustained or non-sustained ventricular tachy-arrhythmia. These findings were supported in a randomized study in 34 MI patients. IC delivery of MSCs was safe and

Table 8.7 Clinical studies with mesenchymal stromal cell (MSC) treatment of patients with ischemic heart disease

	Patients	N treated/controls	Cell type	Cell number	Administration	Treatment effect
Chen et al. 2004	MI	34/35	MSC autologous	6×10^{10}	i.c.	LVEF↑
Katritsis et al. 2005	MI	11/11	MSC autologous		i.c.	LV wall motion↑
Chen et al. 2006	MI	22/62	MSC autologous	5×10^{6}	i.c.	LVEF↑ exercise↑
Katritsis et al. 2007	MI	5/0	MSC autologous		i.c.	No arythmia
Mohyeddin-Bonab et al. 2007		8/8	MSC autologous		i.m.	LVEF↑
Hare et al. 2009	MI	39/21	MSC allogeneic	0.5×10^{6}, 1.6×10^{6}, 5×10^{6}	i.v.	Arythmia↓ lung function↑ LVEF↑
Kastrup et al. 2011	CIHD	31/0	MSC autologous	22×10^{6}	i.m.	Exercise↑ LVEF↑ CCS↓ NTG↓ (6 and 12 months)

MI=acute myocardial infarction; *CIHD*=chronic ischemic heart disease; *DCH*=dilated cardiomyopathy; *I*=open randomized study; *II*=open non-randomized study; *III*=randomized double-blind placebo controlled study; *i.c.*=intracoronary; *i.m.*=intramyocardial; *i.v.*=intravenous; *LVEF*=left ventricular ejection fraction; *CCS*=Canadian Cardiovascular Society angina class; *NTG* Nitroglycerin consumption the trial showed improved LV function after 6 months (Chen et al. 2004). Kastrup et al. have investigated the feasibility, safety and efficacy of direct IM injections of autologous VEGF-A_{165} stimulated MSCs in 31 patients with stable CIHD and refractory angina. The study demonstrated that it was possible and safe to culture expand MSCs and stimulate the cells with VEGF-A_{165} and then to inject the cells directly into the ischemic myocardium of the patients. There was a gradually improvement in the patient's symptoms, exercise capacity, CCS class, angina attacks, nitroglycerine consumption and LVEF from baseline to 3, 6 and 12 months follow-up (Friis et al. 2011; Haack-Sorensen et al. 2013).

Allogeneic bone marrow derived MSCs treatment has been evaluated in patients with MI in a randomized double-blind placebo-controlled study (Hare et al. 2009). The study demonstrated that IV delivery of allogeneic MSCs was safe with identical adverse event rates between MSCs and placebo treated patients. However, MSCs treated patients had reduced ventricular tachycardia episodes, global symptom score and LVEF.

It can be concluded, that both autologous and allogeneic MSC treatment has demonstrated promising clinical results in small trials in patients with MI or CIHD.

Conclusions

Within the last decade increasing evidence has accumulated about the safety and efficacy of both gene and stem cell therapy. The treatments have generally been well tolerated by the patients without major safety concerns. Moreover, the delivery methods also are able to deliver the genes or cells to the target area in the heart without any side-effects (Baldazzi et al. 2008). Therefore, it can be concluded, that until now no serious adverse events has been problematic for gene or stem cell therapy although several different growth factor or stem cell lines have been tested in clinical trials in patients with MI, CIHD and IHF. However, safety issues should still have a high priority in future trials to avoid unexpected side-effects of the gene and stem cell therapy.

The conducted clinical vascular growth factors gene trials in patients with IHD have not demonstrated any significant clinical effect or increased myocardial perfusion. The question is now how to improve the efficacy of the gene therapy. Should larger doses or a combination of growth factors, different vectors and gene delivery methods or combined stem cell and gene therapy be used?

Clinical studies on stem cell therapy for cardiac regeneration in patients with acute and chronic ischemic heart disease have in some but not all studies shown significant improvements in ventricular pump function, ventricular remodeling, myocardial perfusion, exercise capacity and clinical symptoms, compared to conventionally treated control/placebo groups. Moreover, studies indicate a long-term effect of the stem cell treatment on left ventricular function and even on mortality. A recent meta-analysis of 50 clinical trials with stem cell treatment of patients with ischemic heart indicated that there was an improvement in left ventricular function and remodeling and patient-important clinical outcomes as all-cause and cardiac deaths and myocardial infarction (Jeevanantham et al. 2012). The benefit was seen independent of whether it were patients with acute myocardial infarction or chronic myocardial ischemia. The mechanisms behind the potential regenerative capacity of stem cells could be replacement of the myocardium or blood vessels by trans-differentiation or more likely an effect of the cytokines produced by the stem cells on resident cells within the heart.

Different stem cells have been evaluated in different patient populations and the results have not been clear. However, we need to consider that not all stem cells are alike and focus on defined stem cell populations in the future. Presently, there is no knowledge about whether one kind of cell is superior to other when it comes to regenerative capacity, but future trials will probably discover that some stem cells can regenerate and others not. Moreover, we also have to look at the patient populations and the variations within these. Some stem cells might work better in patients with CIHD compared to IHF. In addition many of the patients have co-morbidities such as diabetes, hypercholesterolemia, pulmonary disease etc., which could alter the regenerative capacity of the stem cells. When this information is available it will be possible to move forward for a more individual and personalized gene and/or stem cell treatment strategy in patients with IHD.

There are still many unanswered questions regarding optimal cell type and number, cell source and delivery methods. Some of these questions will be answered in the ongoing or planned larger clinical double-blinded placebo-controlled gene and stem cell trials. The results of these trials will hopefully clarify to which extend genes, stem cells or a combination of genes and stem cells will be a potential new treatment modality in patients with ischemic heart disease.

Acknowledgements This work was supported by a grant of the European Commission (Health-F2-2009-222995, INELPY/Infarct Cell Therapy)

References

Ahmadi H, Baharvand H, Ashtiani SK, Soleimani M, Sadeghian H, Ardekani JM, Mehrjerdi NZ, Kouhkan A, Namiri M, Madani-Civi M, Fattahi F, Shahverdi A, Dizaji AV (2007) Safety analysis and improved cardiac function following local autologous transplantation of CD133(+) enriched bone marrow cells after myocardial infarction. Curr Neurovasc Res 4(3):153–160

Ahn A, Frishman WH, Gutwein A, Passeri J, Nelson M (2008a) Therapeutic angiogenesis: a new treatment approach for ischemic heart disease–part I. Cardiol Rev 16(4):163–171. doi:10.1097/CRD.0b013e3181620e3b

Ahn A, Frishman WH, Gutwein A, Passeri J, Nelson M (2008b) Therapeutic angiogenesis: a new treatment approach for ischemic heart disease–Part II. Cardiol Rev 16(5):219–229. doi:10.1097/CRD.0b013e3181620e50

Assmus B, Schachinger V, Teupe C, Britten M, Lehmann R, Dobert N, Grunwald F, Aicher A, Urbich C, Martin H, Hoelzer D, Dimmeler S, Zeiher AM (2002) Transplantation of progenitor cells and regeneration enhancement in Acute Myocardial Infarction (TOPCARE-AMI). Circulation 106(24):3009–3017

Assmus B, Honold J, Schächinger V, Britten MB, Fischer-Rasokat U, Lehmann R, Teupe C, Pistorius K, Martin H, Abolmaali ND, Tonn T, Dimmeler S, Zeiher AM (2006) Transcoronary transplantation of progenitor cells after myocardial infarction. N Engl J Med 355(12):1222–1232

Baldazzi F, Jorgensen E, Ripa RS, Kastrup J (2008) Release of biomarkers of myocardial damage after direct intramyocardial injection of genes and stem cells via the percutaneous transluminal route. Eur Heart J 29(15):1819–1826. doi:10.1093/eurheartj/ehn233

Beeres SL, Bax JJ, Kaandorp TA, Zeppenfeld K, Lamb HJ, Dibbets-Schneider P, Stokkel MP, Fibbe WE, de Roos A, van der Wall EE, Schalij MJ, Atsma DE (2006) Usefulness of intramyocardial injection of autologous bone marrow-derived mononuclear cells in patients with severe angina pectoris and stress-induced myocardial ischemia. Am J Cardiol 97(9):1326–1331. doi:10.1016/j.amjcard.2005.11.068

Beeres SL, Bax JJ, Dibbets-Schneider P, Stokkel MP, Fibbe WE, van der Wall EE, Schalij MJ, Atsma DE (2007) Intramyocardial injection of autologous bone marrow mononuclear cells in patients with chronic myocardial infarction and severe left ventricular dysfunction. Am J Cardiol 100(7):1094–1098. doi:10.1016/j.amjcard.2007.04.056

Beeres SL, Lamb HJ, Roes SD, Holman ER, Kaandorp TA, Fibbe WE, de Roos A, van der Wall EE, Schalij MJ, Bax JJ, Atsma DE (2008) Effect of intramyocardial bone marrow cell injection on diastolic function in patients with chronic myocardial ischemia. J Magn Reson Imaging 27(5):992–997. doi:10.1002/jmri.21081

Beitnes JO, Gjesdal O, Lunde K, Solheim S, Edvardsen T, Arnesen H, Forfang K, Aakhus S (2011) Left ventricular systolic and diastolic function improve after acute myocardial infarction treated with acute percutaneous coronary intervention, but are not influenced by intracoronary injection of autologous mononuclear bone marrow cells: a 3 year serial echocardiographic substudy of the randomized-controlled ASTAMI study. European journal of echocardiography: the journal of the Working Group on Echocardiography of the European Society of. Cardiology 12(2):98–106. doi:10.1093/ejechocard/jeq116

Chen SL, Fang WW, Ye F, Liu YH, Qian J, Shan SJ, Zhang JJ, Chunhua RZ, Liao LM, Lin S, Sun JP (2004) Effect on left ventricular function of intracoronary transplantation of autologous bone marrow mesenchymal stem cell in patients with acute myocardial infarction. Am J Cardiol 94(1):92–95. doi:10.1016/j.amjcard.2004.03.034

Chen S, Liu Z, Tian N, Zhang J, Yei F, Duan B, Zhu Z, Lin S, Kwan TW (2006) Intracoronary transplantation of autologous bone marrow mesenchymal stem cells for ischemic cardiomyopathy due to isolated chronic occluded left anterior descending artery. J Invasive Cardiol 18(11):552–556

Diederichsen AC, Moller JE, Thayssen P, Junker AB, Videbaek L, Saekmose SG, Barington T, Kristiansen M, Kassem M (2008) Effect of repeated intracoronary injection of bone marrow cells in patients with ischaemic heart failure the Danish stem cell study–congestive heart failure trial (DanCell-CHF). Eur J Heart Fail 10(7):661–667. doi:10.1016/j.ejheart.2008.05.010

Erbs S, Linke A, Adams V, Lenk K, Thiele H, Diederich KW, Emmrich F, Kluge R, Kendziorra K, Sabri O, Schuler G, Hambrecht R (2005) Transplantation of blood-derived progenitor cells after recanalization of chronic coronary artery occlusion: first randomized and placebo-controlled study. Circ Res 97(8):756–762. doi:10.1161/01.RES.0000185811.71306.8b

Friis T, Haack-Sorensen M, Mathiasen AB, Ripa RS, Kristoffersen US, Jorgensen E, Hansen L, Bindslev L, Kjaer A, Hesse B, Dickmeiss E, Kastrup J (2011) Mesenchymal stromal cell derived endothelial progenitor treatment in patients with refractory angina. Scand Cardiovasc J 45(3):161–168. doi:10.3109/14017431.2011.569571

Fuchs S, Dib N, Cohen BM, Okubagzi P, Diethrich EB, Campbell A, Macko J, Kessler PD, Rasmussen HS, Epstein SE, Kornowski R (2006) A randomized, double-blind, placebo-controlled, multicenter, pilot study of the safety and feasibility of catheter-based intramyocardial injection of AdVEGF121 in patients with refractory advanced coronary artery disease. Catheter Cardiovasc Interv 68(3):372–378. doi:10.1002/ccd.20859

Grines CL, Watkins MW, Helmer G, Penny W, Brinker J, Marmur JD, West A, Rade JJ, Marrott P, Hammond HK, Engler RL (2002) Angiogenic Gene Therapy (AGENT) trial in patients with stable angina pectoris. Circulation 105(11):1291–1297

Grines CL, Watkins MW, Mahmarian JJ, Iskandrian AE, Rade JJ, Marrott P, Pratt C, Kleiman N (2003) A randomized, double-blind, placebo-controlled trial of Ad5FGF-4 gene therapy and its effect on myocardial perfusion in patients with stable angina. J Am Coll Cardiol 42(8):1339–1347

Gyongyosi M, Khorsand A, Zamini S, Sperker W, Strehblow C, Kastrup J, Jorgensen E, Hesse B, Tagil K, Botker HE, Ruzyllo W, Teresinska A, Dudek D, Hubalewska A, Ruck A, Nielsen SS, Graf S, Mundigler G, Novak J, Sochor H, Maurer G, Glogar D, Sylven C (2005) NOGA-guided analysis of regional myocardial perfusion abnormalities treated with intramyocardial injections of plasmid encoding vascular endothelial growth factor A-165 in patients with chronic myocardial ischemia: subanalysis of the EUROINJECT-ONE multicenter double-blind randomized study. Circulation 112(9 Suppl):I157–165. doi:10.1161/01.CIRCULATIONAHA.105.525782

Gyongyosi M, Lang I, Dettke M, Beran G, Graf S, Sochor H, Nyolczas N, Charwat S, Hemetsberger R, Christ G, Edes I, Balogh L, Krause KT, Jaquet K, Kuck KH, Benedek I, Hintea T, Kiss R, Preda I, Kotevski V, Pejkov H, Zamini S, Khorsand A, Sodeck G, Kaider A, Maurer G, Glogar D (2009) Combined delivery approach of bone marrow mononuclear stem cells early and late after myocardial infarction: the MYSTAR prospective, randomized study. Nat Clin Pract Cardiovasc Med 6(1):70–81. doi:10.1038/ncpcardio1388

Haack-Sorensen M, Friis T, Mathiasen AB, Jorgensen E, Hansen L, Dickmeiss E, Ekblond A, Kastrup J (2013) Direct intramyocardial mesenchymal stromal cell injections in patients with severe refractory angina: one-year follow-up. Cell Transplant 22(3):521–528. doi:10.3727/096368912X636830

Hare JM, Traverse JH, Henry TD, Dib N, Strumpf RK, Schulman SP, Gerstenblith G, DeMaria AN, Denktas AE, Gammon RS, Hermiller JB Jr, Reisman MA, Schaer GL, Sherman W (2009) A randomized, double-blind, placebo-controlled, dose-escalation study of intravenous adult human mesenchymal stem cells (prochymal) after acute myocardial infarction. J Am Coll Cardiol 54(24):2277–2286. doi:10.1016/j.jacc.2009.06.055

Hedman M, Hartikainen J, Syvanne M, Stjernvall J, Hedman A, Kivela A, Vanninen E, Mussalo H, Kauppila E, Simula S, Narvanen O, Rantala A, Peuhkurinen K, Nieminen MS, Laakso M, Yla-Herttuala S (2003) Safety and feasibility of catheter-based local intracoronary vascular endothelial growth factor gene transfer in the prevention of postangioplasty and in-stent restenosis and in the treatment of chronic myocardial ischemia: phase II results of the Kuopio Angiogenesis Trial (KAT). Circulation 107(21):2677–2683. doi:10.1161/01.CIR.0000070540.80780.92

Hendrikx M, Hensen K, Clijsters C, Jongen H, Koninckx R, Bijnens E, Ingels M, Jacobs A, Geukens R, Dendale P, Vijgen J, Dilling D, Steels P, Mees U, Rummens JL (2006) Recovery of regional but not global contractile function by the direct intramyocardial autologous bone marrow transplantation: results from a randomized controlled clinical trial. Circulation 114 (1 Suppl):I101–107

Henry TD, Grines CL, Watkins MW, Dib N, Barbeau G, Moreadith R, Andrasfay T, Engler RL (2007) Effects of Ad5FGF-4 in patients with angina: an analysis of pooled data from the AGENT-3 and AGENT-4 trials. J Am Coll Cardiol 50(11):1038–1046. doi:10.1016/j.jacc.2007.06.010

Hinkel R, Trenkwalder T, Kupatt C (2011) Gene therapy for ischemic heart disease. Expert Opin Biol Ther 11(6):723–737. doi:10.1517/14712598.2011.570749

Hirsch A, Nijveldt R, van der Vleuten PA, Tijssen JG, van der Giessen WJ, Tio RA, Waltenberger J, ten Berg JM, Doevendans PA, Aengevaeren WR, Zwaginga JJ, Biemond BJ, van Rossum AC, Piek JJ, Zijlstra F (2011) Intracoronary infusion of mononuclear cells from bone marrow or peripheral blood compared with standard therapy in patients after acute myocardial infarction treated by primary percutaneous coronary intervention: results of the randomized controlled HEBE trial. Eur Heart J 32(14):1736–1747. doi:10.1093/eurheartj/ehq449

Janssens S, Dubois C, Bogaert J, Theunissen K, Deroose C, Desmet W, Kalantzi M, Herbots L, Sinnaeve P, Dens J, Maertens J, Rademakers F, Dymarkowski S, Gheysens O, Van Cleemput J, Bormans G, Nuyts J, Belmans A, Mortelmans L, Boogaerts M, Van de Werf F (2006) Autologous bone marrow-derived stem-cell transfer in patients with ST-segment elevation myocardial infarction: double-blind, randomised controlled trial. Lancet 367 (9505):113–121. doi:10.1016/S0140–6736(05)67861–0

Jeevanantham V, Butler M, Saad A, Abdel-Latif A, Zuba-Surma EK, Dawn B (2012) Adult bone marrow cell therapy improves survival and induces long-term improvement in cardiac parameters: a systematic review and meta-analysis. Circulation 126(5):551–568

Kang HJ, Lee HY, Na SH, Chang SA, Park KW, Kim HK, Kim SY, Chang HJ, Lee W, Kang WJ, Koo BK, Kim YJ, Lee DS, Sohn DW, Han KS, Oh BH, Park YB, Kim HS (2006) Differential effect of intracoronary infusion of mobilized peripheral blood stem cells by granulocyte colony-stimulating factor on left ventricular function and remodeling in patients with acute myocardial infarction versus old myocardial infarction: the MAGIC Cell-3-DES randomized, controlled trial. Circulation 114(1 Suppl):I145–151. doi:10.1161/CIRCULATIONAHA.105.001107

Kastrup J (2010) Gene therapy and angiogenesis in patients with coronary artery disease. Expert Rev Cardiovasc Ther 8(8):1127–1138. doi:10.1586/erc.10.95

Kastrup J (2011) Stem cells therapy for cardiovascular repair in ischemic heart disease: how to predict and secure optimal outcome? EPMA J 2(1):107–117. doi:10.1007/s13167-011-0062-5

Kastrup J, Jorgensen E, Ruck A, Tagil K, Glogar D, Ruzyllo W, Botker HE, Dudek D, Drvota V, Hesse B, Thuesen L, Blomberg P, Gyongyosi M, Sylven C (2005) Direct intramyocardial plasmid vascular endothelial growth factor-A165 gene therapy in patients with stable severe angina pectoris A randomized double-blind placebo-controlled study: the Euroinject One trial. J Am Coll Cardiol 45(7):982–988. doi:10.1016/j.jacc.2004.12.068

Kastrup J, Jorgensen E, Fuchs S, Nikol S, Botker HE, Gyongyosi M, Glogar D, Kornowski R (2011) A randomised, double-blind, placebo-controlled, multicentre study of the safety and efficacy of BIOBYPASS (AdGVVEGF121.10NH) gene therapy in patients with refractory advanced coronary artery disease: the NOVA trial. EuroIntervention 6(7):813–818. doi:10.4244/EIJV6I7A140

Katritsis DG, Sotiropoulou PA, Karvouni E, Karabinos I, Korovesis S, Perez SA, Voridis EM, Papamichail M (2005) Transcoronary transplantation of autologous mesenchymal stem cells and endothelial progenitors into infarcted human myocardium. Catheter Cardiovasc Interv 65(3):321–329. doi:10.1002/ccd.20406

Katritsis DG, Sotiropoulou P, Giazitzoglou E, Karvouni E, Papamichail M (2007) Electrophysiological effects of intracoronary transplantation of autologous mesenchymal and endothelial progenitor cells. Europace 9(3):167–171. doi:10.1093/europace/eul184

Kukula K, Chojnowska L, Dabrowski M, Witkowski A, Chmielak Z, Skwarek M, Kadziela J, Teresinska A, Malecki M, Janik P, Lewandowski Z, Klopotowski M, Wnuk J, Ruzyllo W (2011) Intramyocardial plasmid-encoding human vascular endothelial growth factor A165/basic fibroblast growth factor therapy using percutaneous transcatheter approach in patients with refractory coronary artery disease (VIF-CAD). Am Heart J 161(3):581–589. doi:10.1016/j.ahj.2010.11.023

Laitinen M, Hartikainen J, Hiltunen MO, Eranen J, Kiviniemi M, Narvanen O, Makinen K, Manninen H, Syvanne M, Martin JF, Laakso M, Yla-Herttuala S (2000) Catheter-mediated vascular endothelial growth factor gene transfer to human coronary arteries after angioplasty. Hum Gene Ther 11(2):263–270. doi:10.1089/10430340050016003

Leistner DM, Fischer-Rasokat U, Honold J, Seeger FH, Schachinger V, Lehmann R, Martin H, Burck I, Urbich C, Dimmeler S, Zeiher AM, Assmus B (2011) Transplantation of progenitor cells and regeneration enhancement in acute myocardial infarction (TOPCARE-AMI): final 5-year results suggest long-term safety and efficacy. Clin Res Cardiol 100(10):925–934. doi:10.1007/s00392–011-0327-y

Losordo DW, Vale PR, Hendel RC, Milliken CE, Fortuin FD, Cummings N, Schatz RA, Asahara T, Isner JM, Kuntz RE (2002) Phase 1/2 placebo-controlled, double-blind, dose-escalating trial of myocardial vascular endothelial growth factor 2 gene transfer by catheter delivery in patients with chronic myocardial ischemia. Circulation 105(17):2012–2018

Losordo DW, Schatz RA, White CJ, Udelson JE, Veereshwarayya V, Durgin M, Poh KK, Weinstein R, Kearney M, Chaudhry M, Burg A, Eaton L, Heyd L, Thorne T, Shturman L, Hoffmeister P, Story K, Zak V, Dowling D, Traverse JH, Olson RE, Flanagan J, Sodano D, Murayama T, Kawamoto A, Kusano KF, Wollins J, Welt F, Shah P, Soukas P, Asahara T, Henry TD (2007) Intramyocardial transplantation of autologous CD34+ stem cells for intractable angina: a phase I/IIa double-blind, randomized controlled trial. Circulation 115(25):3165–3172. doi:10.1161/CIRCULATIONAHA.106.687376

Lunde K, Solheim S, Aakhus S, Arnesen H, Abdelnoor M, Egeland T, Endresen K, Ilebekk A, Mangschau A, Fjeld JG, Smith HJ, Taraldsrud E, Grogaard HK, Bjornerheim R, Brekke M, Muller C, Hopp E, Ragnarsson A, Brinchmann JE, Forfang K (2006) Intracoronary injection of mononuclear bone marrow cells in acute myocardial infarction. New Engl J Med 355(12):1199–1209. doi:10.1056/NEJMoa055706

Mathiasen AB, Haack-Sorensen M, Kastrup J (2009) Mesenchymal stromal cells for cardiovascular repair: current status and future challenges. Future Cardiol 5(6):605–617. doi:10.2217/fca.09.42

Menasche P, Alfieri O, Janssens S, McKenna W, Reichenspurner H, Trinquart L, Vilquin JT, Marolleau JP, Seymour B, Larghero J, Lake S, Chatellier G, Solomon S, Desnos M, Hagege AA (2008) The Myoblast Autologous Grafting in Ischemic Cardiomyopathy (MAGIC) trial: first randomized placebo-controlled study of myoblast transplantation. Circulation 117(9):1189–1200. doi:10.1161/CIRCULATIONAHA.107.734103

Meyer GP, Wollert KC, Lotz J, Steffens J, Lippolt P, Fichtner S, Hecker H, Schaefer A, Arseniev L, Hertenstein B, Ganser A, Drexler H (2006) Intracoronary bone marrow cell transfer after myocardial infarction: eighteen months' follow-up data from the randomized, controlled BOOST (BOne marrOw transfer to enhance ST-elevation infarct regeneration) trial. Circulation 113(10):1287–1294. doi:10.1161/CIRCULATIONAHA.105.575118

Meyer GP, Wollert KC, Lotz J, Pirr J, Rager U, Lippolt P, Hahn A, Fichtner S, Schaefer A, Arseniev L, Ganser A, Drexler H (2009) Intracoronary bone marrow cell transfer after myocardial infarction: 5-year follow-up from the randomized-controlled BOOST trial. Eur Heart J 30(24):2978–2984. doi:10.1093/eurheartj/ehp374

Perin EC, Dohmann HF, Borojevic R, Silva SA, Sousa AL, Mesquita CT, Rossi MI, Carvalho AC, Dutra HS, Dohmann HJ, Silva GV, Belem L, Vivacqua R, Rangel FO, Esporcatte R, Geng YJ, Vaughn WK, Assad JA, Mesquita ET, Willerson JT (2003) Transendocardial, autologous bone marrow cell transplantation for severe, chronic ischemic heart failure. Circulation 107(18):2294–2302. doi:10.1161/01.CIR.0000070596.30552.8B

Perin EC, Silva GV, Henry TD, Cabreira-Hansen MG, Moore WH, Coulter SA, Herlihy JP, Fernandes MR, Cheong BY, Flamm SD, Traverse JH, Zheng Y, Smith D, Shaw S, Westbrook L, Olson R, Patel D, Gahremanpour A, Canales J, Vaughn WK, Willerson JT (2011) A randomized study of transendocardial injection of autologous bone marrow mononuclear cells and cell function analysis in ischemic heart failure (FOCUS-HF). Am Heart J 161(6):1078–1087 e1073. doi:10.1016/j.ahj.2011.01.028

Pittenger MF, Martin BJ (2004) Mesenchymal stem cells and their potential as cardiac therapeutics. Circ Res 95(1):9–20. doi:10.1161/01.RES.0000135902.99383.6f

Ripa RS, Wang Y, Jorgensen E, Johnsen HE, Hesse B, Kastrup J (2006) Intramyocardial injection of vascular endothelial growth factor-A165 plasmid followed by granulocyte-colony stimulating factor to induce angiogenesis in patients with severe chronic ischaemic heart disease. Eur Heart J 27(15):1785–1792. doi:10.1093/eurheartj/ehl117

Roncalli J, Mouquet F, Piot C, Trochu JN, Le Corvoisier P, Neuder Y, Le Tourneau T, Agostini D, Gaxotte V, Sportouch C, Galinier M, Crochet D, Teiger E, Richard MJ, Polge AS, Beregi JP, Manrique A, Carrie D, Susen S, Klein B, Parini A, Lamirault G, Croisille P, Rouard H, Bourin P, Nguyen JM, Delasalle B, Vanzetto G, Van Belle E, Lemarchand P (2011) Intracoronary autologous mononucleated bone marrow cell infusion for acute myocardial infarction: results of the randomized multicenter BONAMI trial. Eur Heart J 32(14):1748–1757. doi:10.1093/eurheartj/ehq455

Schachinger V, Assmus B, Britten MB, Honold J, Lehmann R, Teupe C, Abolmaali ND, Vogl TJ, Hofmann WK, Martin H, Dimmeler S, Zeiher AM (2004) Transplantation of progenitor cells and regeneration enhancement in acute myocardial infarction: final one-year results of the TOPCARE-AMI Trial. J Am Coll Cardiol 44(8):1690–1699. doi:10.1016/j.jacc.2004.08.014

Schachinger V, Erbs S, Elsasser A, Haberbosch W, Hambrecht R, Holschermann H, Yu J, Corti R, Mathey DG, Hamm CW, Suselbeck T, Assmus B, Tonn T, Dimmeler S, Zeiher AM (2006a)

Intracoronary bone marrow-derived progenitor cells in acute myocardial infarction. New Engl J Med 355(12):1210–1221. doi:10.1056/NEJMoa060186

Schachinger V, Erbs S, Elsasser A, Haberbosch W, Hambrecht R, Holschermann H, Yu J, Corti R, Mathey DG, Hamm CW, Suselbeck T, Werner N, Haase J, Neuzner J, Germing A, Mark B, Assmus B, Tonn T, Dimmeler S, Zeiher AM (2006b) Improved clinical outcome after intracoronary administration of bone-marrow-derived progenitor cells in acute myocardial infarction: final 1-year results of the REPAIR-AMI trial. Eur Heart J 27(23):2775–2783. doi:10.1093/eurheartj/ehl388

Stewart DJ, Hilton JD, Arnold JM, Gregoire J, Rivard A, Archer SL, Charbonneau F, Cohen E, Curtis M, Buller CE, Mendelsohn FO, Dib N, Page P, Ducas J, Plante S, Sullivan J, Macko J, Rasmussen C, Kessler PD, Rasmussen HS (2006) Angiogenic gene therapy in patients with nonrevascularizable ischemic heart disease: a phase 2 randomized, controlled trial of Ad-VEGF(121) (AdVEGF121) versus maximum medical treatment. Gene Ther 13(21):1503–1511. doi:10.1038/sj.gt.3302802

Stewart DJ, Kutryk MJ, Fitchett D, Freeman M, Camack N, Su Y, Della Siega A, Bilodeau L, Burton JR, Proulx G, Radhakrishnan S (2009) VEGF gene therapy fails to improve perfusion of ischemic myocardium in patients with advanced coronary disease: results of the NORTHERN trial. Molecular therapy: the journal of the American Society of. Gene Ther 17(6):1109–1115. doi:10.1038/mt.2009.70

Strauer BE, Yousef M, Schannwell CM (2010) The acute and long-term effects of intracoronary Stem cell Transplantation in 191 patients with chronic heARt failure: the STAR-heart study. Eur J Heart Fail 12(7):721–729. doi:10.1093/eurjhf/hfq095

Symes JF, Losordo DW, Vale PR, Lathi KG, Esakof DD, Mayskiy M, Isner JM (1999) Gene therapy with vascular endothelial growth factor for inoperable coronary artery disease. Ann Thorac Surg 68(3):830–836; discussion 836–837

Tendera M, Wojakowski W, Ruzyllo W, Chojnowska L, Kepka C, Tracz W, Musialek P, Piwowarska W, Nessler J, Buszman P, Grajek S, Breborowicz P, Majka M, Ratajczak MZ (2009) Intracoronary infusion of bone marrow-derived selected CD34+ CXCR4+ cells and non-selected mononuclear cells in patients with acute STEMI and reduced left ventricular ejection fraction: results of randomized, multicentre myocardial regeneration by Intracoronary Infusion of Selected Population of Stem Cells in Acute Myocardial Infarction (REGENT) Trial. Eur Heart J 30(11):1313–1321. doi:10.1093/eurheartj/ehp073

Tio RA, Tan ES, Jessurun GA, Veeger N, Jager PL, Slart RH, de Jong RM, Pruim J, Hospers GA, Willemsen AT, de Jongste MJ, van Boven AJ, van Veldhuisen DJ, Zijlstra F (2004) PET for evaluation of differential myocardial perfusion dynamics after VEGF gene therapy and laser therapy in end-stage coronary artery disease. J Nucl Med 45(9):1437–1443

Traverse JH, Henry TD, Vaughan DE, Ellis SG, Pepine CJ, Willerson JT, Zhao DX, Simpson LM, Penn MS, Byrne BJ, Perin EC, Gee AP, Hatzopoulos AK, McKenna DH, Forder JR, Taylor DA, Cogle CR, Baraniuk S, Olson RE, Jorgenson BC, Sayre SL, Vojvodic RW, Gordon DJ, Skarlatos SI, Moyè LA, Simari RD (2010) Cardiovascular cell therapy research network. LateTIME: a phase-II, randomized, double-blinded, placebo-controlled, pilot trial evaluating the safety and effect of administration of bone marrow mononuclear cells 2 to 3 weeks after acute myocardial infarction. Tex Heart Inst J 37(4):412–420

Tse HF, Siu CW, Zhu SG, Songyan L, Zhang QY, Lai WH, Kwong YL, Nicholls J, Lau CP(2007) Paracrine effects of direct intramyocardial implantation of bone marrow derived cells to enhance neovascularization in chronic ischaemic myocardium. Eur J Heart Fail 9(8):747–753

van Ramshorst J, Bax JJ, Beeres SL, Dibbets-Schneider P, Roes SD, Stokkel MP, de Roos A, Fibbe WE, Zwaginga JJ, Boersma E, Schalij MJ, Atsma DE (2009) Intramyocardial bone marrow cell injection for chronic myocardial ischemia: a randomized controlled trial. JAMA 301(19):1997–2004. doi:10.1001/jama.2009.685

Wang Y, Tagil K, Ripa RS, Nilsson JC, Carstensen S, Jorgensen E, Sondergaard L, Hesse B, Johnsen HE, Kastrup J (2005) Effect of mobilization of bone marrow stem cells by granulocyte colony stimulating factor on clinical symptoms, left ventricular perfusion and function in patients with severe chronic ischemic heart disease. Int J Cardiol 100(3):477–483. doi:10.1016/j.ijcard.2004.12.006

Willems E, Lanier M, Forte E, Lo F, Cashman J, Mercola M (2011) A chemical biology approach to myocardial regeneration. J Cardiovasc Transl Res 4(3):340–350. doi:10.1007/s12265–011-9270–6

Wohrle J, Merkle N, Mailander V, Nusser T, Schauwecker P, von Scheidt F, Schwarz K, Bommer M, Wiesneth M, Schrezenmeier H, Hombach V (2010) Results of intracoronary stem cell therapy after acute myocardial infarction. Am J Cardiol 105(6):804–812. doi:10.1016/j.amjcard.2009.10.060

Wollert KC, Meyer GP, Lotz J, Ringes-Lichtenberg S, Lippolt P, Breidenbach C, Fichtner S, Korte T, Hornig B, Messinger D, Arseniev L, Hertenstein B, Ganser A, Drexler H (2004) Intracoronary autologous bone-marrow cell transfer after myocardial infarction: the BOOST randomised controlled clinical trial. Lancet 364 (9429):141–148. doi:10.1016/S0140–6736(04)16626–9

Yousef M, Schannwell CM, Kostering M, Zeus T, Brehm M, Strauer BE (2009) The BALANCE Study: clinical benefit and long-term outcome after intracoronary autologous bone marrow cell transplantation in patients with acute myocardial infarction. J Am Coll Cardiol 53(24):2262–2269. doi:10.1016/j.jacc.2009.02.051

Index

J. Hescheler, E. Hofer (eds.), *Adult and Pluripotent Stem Cells*,
DOI 10.1007/978-94-017-8657-7, © Springer Science+Business Media Dordrecht 2014

MIX
Papier aus verantwortungsvollen Quellen
Paper from responsible sources
FSC® C105338

If you have any concerns about our products, you can contact us on
ProductSafety@springernature.com

In case Publisher is established outside the EU, the EU authorized representative is:
Springer Nature Customer Service Center GmbH
Europaplatz 3, 69115 Heidelberg, Germany

Printed by Libri Plureos GmbH
in Hamburg, Germany